United States
Environmental Protection
Agency

Climate Ready Water Utilities
Adaptation Strategies Guide
for Water Utilities

CLIMATE READY
WATER UTILITIES
♼EPA

Office of Water (4608-T)   EPA 817-K-15-001   February 2015

Finally, **Sustainability Briefs** describe how sustainable strategies, specifically Green Infrastructure, Energy Management and Water Demand Management activities, can be used in conjunction with adaptation planning. These Sustainability Briefs can be accessed at any point in the Guide, and provide information on the benefits of sustainable practices, and how utilities can begin implementing some of these practices. If your utility is already pursuing Green Infrastructure, Energy Management or Water Demand Management activities, these briefs may help you to identify opportunities for coordinating adaptation strategies with these existing sustainable initiatives. If your utility has not yet considered sustainable practices, you can learn more about these options by reviewing the Sustainability Briefs.

If you have any questions about or feedback on the Adaptation Strategies Guide, or would like to suggest new material (e.g., examples) to include, please email CRWUhelp@epa.gov.

HOW TO USE THIS GUIDE

You can navigate this Guide as if it were a website. Instructions indicating clickable links can be found in the Table of Contents, the last section of the Introduction and within all of the briefs. Many of the links are represented with an icon or picture, while others are hyperlinked and displayed with underlined text (e.g., *Worksheet for Adaptation Planning*). Clicking on the [Return to Introduction] button will bring you back to the last section of the Introduction where you can select any of the briefs in the Guide. **From anywhere in the Guide, you can also return to a previously viewed page by pressing the ALT key with the left arrow key.**

MWRA has continued to include sea-level rise consideration into adaptation planning, facility designs and rehabilitation projects, and has made significant investments in climate change mitigation through energy efficiency and green energy development. About half of the energy used by the Deer Island plant is produced on-site through the use of renewable sources, including digester gas, wind, solar and hydroelectric generation. In addition, the use of hydroelectric generation within the drinking water systems produces enough green power to offset over half of the energy used at the 405 million gallon per day Carroll Water Treatment Plant.

EXAMPLE 2

The Southern Monmouth Regional Sewer Authority (SMRSA) provides wastewater service for many coastal New Jersey communities. SMRSA is at risk to impacts from coastal storms and future sea-level rise, and was severely impacted by Hurricane Irene in 2011 and Superstorm Sandy in 2012. To address impacts related to coastal storms, SMRSA has implemented a state-of-the-art solution of constructing mobile pump station enclosures.

The mobile pump stations are designed to house the primary electrical equipment and controls in a mobile enclosure that would elevate the equipment above the level of flood damage. When an approaching coastal storm has the potential to damage the pump station, the enclosure can be removed from the site and transported to an area of higher elevation. An expendable portable generator and transfer switch are transported to the pump site to operate the station if utility power is lost. A secondary sacrificial electrical and control system is permanently mounted at the site and will operate the pumps on utility or generator power. Once the storm subsides, the enclosure can be moved back to the station and all electrical equipment put back online. The first of these units was constructed in 2011 to protect a pump station that had been damaged and partially flooded due to previous coastal storms. This station provided protection during Hurricane Irene and Superstorm Sandy, allowing the utility to continue providing wastewater services.

The use of these mobile stations minimizes damage to the pumping station's electrical equipment, significantly reducing any downtime of the station and allowing the utility to return the station to normal operation within hours of the passing of the storm. Traditionally, if electrical and control equipment of a pumping station were damaged, the station could potentially be out of service for weeks until the equipment was replaced, resulting in significant financial and environmental costs. Because of their effectiveness and the likelihood of additional storms of increasing magnitudes, SMRSA plans to have three additional mobile pumping stations constructed by the summer of 2015 (SMRSA 2012).

 United States
Environmental Protection
Agency

CLIMATE READY
WATER UTILITIES
⚘EPA

Group: ECOSYSTEM CHANGES (DW/WW)

Return to Introduction

The impacts to water utilities from ecosystem changes associated with climate change may be driven or forced by loss of coastal systems, increases in wildfires and altered vegetation. Clicking on either the drinking water or wastewater icon next to each impact will bring you to that particular Strategy Brief. Clicking on the Green Infrastructure icon will bring you to that Sustainability Brief.

Loss of Coastal Landforms and Wetlands DW WW

Sea-level rise and increasing frequency of damaging tropical storms can lead to losses of coastal and stream ecosystems. Loss of coastal wetlands can reduce the buffer against coastal storms, which may damage coastal treatment plants and infrastructure and lead to service disruptions. Review this brief to learn more about how the UK Environment Agency implemented a strategic flood risk management plan for the Thames Estuary, including suggestions for exploring habitat restoration and constructing flood barriers.

Increased Fire Risk and Altered Vegetation DW WW

Changes in climate are likely to disturb the ecosystem and alter the diversity of vegetation. These changes, coupled with potential droughts or changes in evaporation and soil-water retention, may lead to increased risks of wildfire. In addition to potential degradation of water supply, fires present a direct risk to property and infrastructure. Runoff and flash floods from burned areas can increase sedimentation in reservoirs, reducing their capacity and effective service lifespan. In reservoirs, increased pollutant loads, such as heavy metals and nutrients, could result in higher turbidity, algal blooms and subsequent higher treatment costs. Review this brief to learn more about how Denver Water addressed issues related to flash flooding following wildfire events through increased water treatment and infrastructure updates and by practicing fire management activities.

ADAPTATION OPTIONS

Click to left of name to check off options for consideration; $'s (**$-$$$**) indicate relative costs
Click name of any option to review more information in the Glossary
⭐ **No Regrets options** - actions that would provide benefits to the utility under current climate conditions as well as any future changes in climate. For more information on No Regrets options, see Page 11 in the Introduction.
Click on the 💡, 💧 or ♻ icon to review the relevant Sustainability Brief.

✓	PLANNING	COST
	Study response of nearby wetlands to storm surge events.	$
	Update fire models and fire management plans to incorporate any changes in fire frequency, magnitude and extent due to projected future climate conditions.	$-$$
	Conduct sea-level rise and storm surge modeling. Incorporate resulting inundation mapping and frequency estimates into land use and facility planning.	$
⭐	Develop models to understand potential water quality changes (e.g., increased turbidity) and costs of resultant changes in treatment.	$
💡	Plan for alternative power supplies to support operations in case of loss of power.	$
	Adopt insurance mechanisms and other financial instruments, such as catastrophe bonds, to protect against financial losses associated with infrastructure losses.	$
⭐	Conduct climate change impacts and adaptation training for personnel.	$
⭐	Develop coastal restoration plans, including consideration of barrier islands, coastal wetlands and dune ecosystems.	$-$$

Continued on page 2

✓	PLANNING	COST
⊛	Ensure that emergency response plans deal with flooding and wildfire and include stakeholder engagement and communication.	$
	Integrate climate-related risks into capital improvement plans, including options that provide resilience against current and potential future sea-level and storm surge risks.	$
⊙	Participate in community planning and regional collaborations related to climate change adaptation.	$ $$
	Implement policies and procedures for post-flood and/or post-fire repairs.	$

✓	OPERATIONAL STRATEGIES	COST
	Practice fire management plans in the watershed, such as mechanical thinning, weed control, selective harvesting, controlled burns and creation of fire breaks.	$-$$
⊛	Monitor and inspect the integrity of existing infrastructure.	$-$$
⊛	Monitor current weather conditions, including precipitation and temperature.	$
⊛	Monitor flood events and drivers that may impact flood and water quality models (e.g., precipitation, catchment runoff, storm intensity, sea level).	$
⊛	Monitor surface water conditions, including streamflow and water quality.	$
	Monitor vegetation changes in watersheds.	$

✓	CAPITAL/INFRASTRUCTURE STRATEGIES	COST
⊛	Acquire and manage coastal ecosystems, such as coastal wetlands, to attenuate storm surge and reduce coastal flooding ("soft protection").	$$$
⊙	Acquire and manage ecosystems, such as forested watersheds, vegetation strips and wetlands, to buffer against floods and sediment and nutrient inflows into source waterways.	$$$
⊛	Set aside land to support future flood-proofing needs (e.g., berms, dikes and retractable gates).	$$$
	Implement or retrofit source control measures that address altered influent flow and quality at treatment plants.	$$-$$$
	Build flood barriers, sea walls, levees and related structures to protect infrastructure.	$$-$$$
⊛ ⊙	Diversify options to complement current water supply, including recycled water, desalination, conjunctive use and stormwater capture.	$$$
⊛	Expand current resources by developing regional water connections to allow for water trading in times of service disruption or shortage.	$$-$$$
⊛	Increase water storage capacity, including silt removal to expand capacity at existing reservoirs and construction of new reservoirs and/or dams.	$$-$$$
⊙	Establish alternative power supplies, potentially through on-site generation, to support operations in case of loss of power.	$-$$
	Relocate facilities (e.g., treatment plants) to higher ground.	$$$
	Implement barriers and aquifer recharge to limit effects of saltwater intrusion. Consider use of reclaimed water to create saltwater intrusion barriers.	$$$
⊛	Increase treatment capabilities to address water quality changes (e.g., increased turbidity or salinity).	$$$

 United States
Environmental Protection
Agency

CLIMATE READY
WATER UTILITIES
♻EPA

LOSS OF COASTAL LANDFORMS / WETLANDS (DW/WW)

Return to Introduction

Global mean sea level has risen by 8 inches since 1880 and is projected to continue to rise another 1 to 4 feet by 2100 (USGCRP 2014). Sea-level rise and increasing frequency of damaging tropical storms can lead to losses of coastal and stream ecosystems. Loss of coastal wetlands can reduce the buffer against coastal storms, leading to damage to coastal treatment plants and infrastructure, such as intake facilities and water conveyance and distribution systems, and may cause disruption of service.

CLIMATE INFORMATION

- Climate change-induced sea-level rise is due to two processes: thermal expansion of the oceans as they warm and melting of glaciers and ice sheets (Antarctica, Greenland) on land. The IPCC Fifth Assessment Report estimates that global mean sea level will rise 0.26 to 0.82 meters (10.2 to 32.3 inches) over the course of the 21st century (IPCC 2012). Other scientists estimate that sea-level rise could reach 6.6 feet by the end of the century (Parris et al. 2012).

- Local observed sea level is due to a combination of factors including changes in global mean sea level, regional differences due to the influence of ocean currents, salinity and other local dynamics such as subsidence, and in some cases, tectonic uplift (common in Alaska). A recent study demonstrates that, over the past 60 years, sea level along the Gulf of Mexico has been rising substantially faster (5 to 10 mm/year) than the global trend (1.7 mm/year) due to land subsidence. Subsidence is also responsible for faster than average sea-level rise in the Mid-Atlantic region. For example, subsidence has increased from 2 to 3 cm in the past 40 years in southern New Jersey due to groundwater withdrawals (Parris et al. 2012). In the Northeast, sea-level rise is expected to exceed the global average by up to 4 inches per century.

- Sea-level rise will cause the level of flooding that occurs during the current 100-year storms to occur more frequently by mid-century (USGCRP 2014). For example, the 1-in-100 year coastal flood event in New York City is expected to occur once in every 15 to 35 years by the end of the century (Horton 2010).

- Sea-level rise is a gradual coastal flooding threat, but it will exacerbate more sudden coastal storm surges during severe storms, including – but not limited to – hurricanes. The intensity, frequency and duration of North Atlantic hurricanes has increased in recent decades, and the intensity of these storms is likely to continue to increase in this century (USGCRP 2014). More intense hurricanes can be expected to lead to increased flooding in coastal and near-coast areas.

ADAPTATION OPTIONS

Click to left of name to check off options for consideration; $'s (**$-$$$**) indicate relative costs
Click name of any option to review more information in the Glossary
⭐ **No Regrets options** - actions that would provide benefits to the utility under current climate conditions as well as any future changes in climate. For more information on No Regrets options, see Page 11 in the Introduction.
Click on the 💡, 💧 or ☕ icon to review the relevant Sustainability Brief.

✔	PLANNING	COST
	Study response of nearby wetlands to storm surge events.	$
	Conduct sea-level rise and storm surge modeling. Incorporate resulting inundation mapping and frequency estimates into land use and facility planning.	$
⭐	Develop models to understand potential water quality changes (e.g., increased turbidity) and costs of resultant changes in treatment.	$
💡	Plan for alternative power supplies to support operations in case of loss of power.	$
	Adopt insurance mechanisms and other financial instruments, such as catastrophe bonds, to protect against financial losses associated with infrastructure losses.	$
⭐	Conduct climate change impacts and adaptation training for personnel.	$

Continued on page 2

✓ PLANNING (continued)	COST
Develop coastal restoration plans, including consideration of barrier islands, coastal wetlands and dune ecosystems.	$-$$
Develop emergency response plans to deal with flooding contingencies and include stakeholder engagement and communication.	$
Integrate climate-related risks into capital improvement plans, including options that provide resilience against current and potential future sea-level and storm surge risks.	$
Participate in community planning and regional collaborations related to climate change adaptation.	$-$$
Implement policies and procedures for post-flood repairs.	$

✓ OPERATIONAL STRATEGIES	COST
Monitor and inspect the integrity of existing infrastructure. (**See example below**)	$-$$
Monitor current weather conditions, including precipitation and temperature.	$
Monitor flood events and drivers that may impact flood and water quality models (e.g., storm intensity, sea level).	$

✓ CAPITAL/INFRASTRUCTURE STRATEGIES	COST
Acquire and manage coastal ecosystems, such as coastal wetlands, to attenuate storm surge and reduce coastal flooding ("soft protection"). (**See example below**)	$$$
Acquire and manage ecosystems, such as forested watersheds, vegetation strips and wetlands, to buffer against floods and sediment and nutrient inflows into source waterways. (**See example below**)	$$$
Set aside land to support future flood-proofing needs (e.g., berms, dikes and retractable gates).	$$$
Build flood barriers, sea walls, levees and related structures to protect infrastructure. (**See example below**)	$$-$$$
Diversify options to complement current water supply, including recycled water, desalination, conjunctive use and stormwater capture.	$$$
Expand current resources by developing regional water connections to allow for water trading in times of service disruption or shortage.	$$-$$$
Establish alternative power supplies, potentially through on-site generation, to support operations in case of loss of power.	$-$$
Relocate facilities (e.g., treatment plants) to higher ground.	$$$
Implement barriers and aquifer recharge to limit effects of saltwater intrusion. Consider use of reclaimed water to create saltwater intrusion barriers.	$$$
Increase treatment capabilities to address water quality changes (e.g., increased turbidity or salinity).	$$$

EXAMPLE

Climate change will increase flood risk on the River Thames in England, from the combined impacts of extreme storms inland and tidal surge and sea-level rise in coastal areas. Sea-level rise may be on the order of 8 to 35 inches by 2100. In some locations, future peak freshwater flows for the Thames could increase by around 40% by 2080. The River Thames is a 350 km river of which about 100 km is tidal; the tidal section includes the portion flowing through London. Alarmingly, an estimated 45,000 properties in the non-tidal section are vulnerable to a 100-year flood event. In the tidal section, more than 500,000 properties are at risk of flooding.

The UK Environment Agency established the Thames Estuary 2100 Plan to develop a strategic flood risk management plan for London and the Thames estuary. The project included consideration of how tidal flood risk was likely to change in response to future changes in climate, land use and demographics, and how the existing flood defense system should be adapted to these potential changes. Eight major flood barriers, along with 36 industrial flood gates, over 400 moveable structures and 330km of walls and embankments currently provide flood protection along the Thames.

The Plan is organized in three phases (2010-2034, 2035-2069 and 2070-2100) which include measures that respond to sea-level rise and the ongoing deterioration and aging of the defense system. The measures include maintaining and repairing existing defenses, raising of defenses, realignment of defenses, new or improved tidal flood barriers and increased emergency planning activities. Habitat restoration is also a component of the Plan; the UK intends to invest in more than 1,200 hectares of new intertidal wetlands habitat in or near the Thames estuary by 2100. Furthermore, freshwater habitat will be created to compensate for any loss in the above intertidal expansion.

The Plan will be updated periodically as new data for sea-level rise, estimated increase in peak surge tide levels and current conditions of each section of the flood defense system become available.

 United States
Environmental Protection
Agency

CLIMATE READY
WATER UTILITIES

INCREASED FIRE RISK & ALTERED VEGETATION (DW/WW) Return to Introduction

Changes in climate are likely to disturb the ecosystem and alter the diversity of vegetation. These changes, coupled with potential droughts or changes in evaporation and soil-water retention, may lead to increased risks of wildfire. Fires present a direct risk to property and infrastructure, in addition to potential degradation of water supply. Runoff and flash floods from burned areas can increase sedimentation in reservoirs, reducing their capacity and effective service lifespan. In reservoirs, increased pollutant loads, such as heavy metals and nutrients, could result in higher turbidity, algal blooms and subsequent higher treatment costs.

CLIMATE INFORMATION

- While the average number of wildfires per year has decreased over time, the frequency of large wildfires and length of fire season have increased substantially since 1985, and the total area burned by wildfires in the continental United States has nearly doubled since 2000 due to extremely dry conditions and land management practices (USGCRP 2014). Projected increases in fire frequency and severity are expected, particularly in the western states and Alaska. Models project a doubling of burned area in the southern Rockies (Litschert et al. 2012) and up to 74% more fires in California (Westerling et al. 2011). This trend is most closely linked with earlier spring snowmelt. Much of this increased fire activity occurred in the mid-elevation forests in the northern Rocky Mountains and Sierra Nevada Mountains (Westerling et al. 2006, CCSP 2008). Earlier snowmelt contributes to fire frequency by increasing the ignition period and decreasing water availability later in the summer, increasing fuel loads.

- Seasonal and multi-year droughts affect wildfire severity (Brown et al. 2008, Littell et al. 2009, Schoennagel 2011, Westerling et al. 2003). Five western states (Arizona, Colorado, Utah, California and New Mexico) have experienced their largest fires on record at least once since 2000. Much of the increase in fires larger than 500 acres occurred in the western U.S., while the area burned in the Southwest increased more than 300% relative to the area burned during the 1970s and early 1980s (Westerling et al. 2006).

- By 2070, the length of the fire season could increase by 2 to 3 weeks in the southwestern U.S. (Barnett et al. 2004). Climate models project that there will be more low humidity days in the western U.S. in the future, allowing for more fire activity. Future snowpacks are also expected to be reduced. In California's Sierra Nevada Mountains, for example, snowpack reductions are projected to range from 25% to 40% by 2050 (Standish-Lee and Lecina 2008). Insect pests, such as bark beetles, are expected to expand their range and exacerbate fire conditions. This increased wildfire activity is not confined to the West; the forest fire seasonal severity rating (related to fire intensity and difficulty of fire control) is projected to increase 10% to 30% in the Southeast and 10% to 20% in the Northeast by 2060 (Flannigan et al. 2000).

- Ecosystems will be altered, such as possible shifts from closed forest to savanna and grassland in the Southeast during this century, due to projected changes in temperature and precipitation. Many tree species will have distributions that are shifted northward and to higher elevations. In general, invasive plants will benefit from increased warming (CCSP 2008, USGCRP 2009). These phenomena will have complex and unpredictable impacts on water availability and runoff in watersheds.

Continued on page 2

ADAPTATION OPTIONS

Click to left of name to check off options for consideration; $'s ($-$$$) indicate relative costs
Click name of any option to review more information in the Glossary
⊛ **No Regrets options** - actions that would provide benefits to the utility under current climate conditions as well as any future changes in climate. For more information on No Regrets options, see Page 11 in the Introduction.
Click on the 💡, 👤 or 🌐 icon to review the relevant Sustainability Brief.

✓	PLANNING	COST
	Update fire models and fire management plans to incorporate any changes in fire frequency, magnitude and extent due to projected future climate conditions.	$-$$
⊛	Develop models to understand potential water quality changes (e.g., increased turbidity) and costs of resultant changes in treatment.	$
💡	Plan for alternative power supplies to support operations in case of loss of power.	$
	Adopt insurance mechanisms and other financial instruments, such as catastrophe bonds, to protect against financial losses associated with infrastructure losses.	$
⊛	Conduct climate change impacts and adaptation training for personnel.	$
⊛	Ensure that emergency response plans deal with flooding and wildfire and include stakeholder engagement and communication.	$
🌐	Participate in community planning and regional collaborations related to climate change adaptation.	$-$$
	Implement policies and procedures for post-flood and/or post-fire repairs.	$

✓	OPERATIONAL STRATEGIES	COST
	Practice fire management plans in the watershed, such as mechanical thinning, weed control, selective harvesting, controlled burns and creation of fire breaks. *(See example below)*	$-$$
⊛	Monitor flood events and drivers that may impact flood and water quality models (e.g., precipitation, catchment runoff). *(See example below)*	$
⊛	Monitor surface water conditions, including streamflow and water quality.	$
	Monitor vegetation changes in watersheds.	$

✓	CAPITAL/INFRASTRUCTURE STRATEGIES	COST
	Implement or retrofit source control measures that address altered influent flow and quality at treatment plants.	$$-$$$
	Build flood barriers, flood control dams, levees and related structures to protect infrastructure.	$$-$$$
⊛ 💧	Diversify options to complement current water supply, including recycled water, desalination, conjunctive use and stormwater capture.	$$$
⊛	Expand current resources by developing regional water connections to allow for water trading in times of service disruption or shortage.	$$-$$$
⊛	Increase water storage capacity, including silt removal to expand capacity at existing reservoirs and construction of new reservoirs and/or dams.	$$-$$$
💡	Establish alternative power supplies, potentially through on-site generation, to support operations in case of loss of power.	$-$$
⊛	Increase treatment capabilities to address water quality changes (e.g., increased turbidity).	$$$

EXAMPLE

A series of natural disasters in Denver Water's primary watershed including the 1996 Buffalo Creek Fire (which burned 11,900 acres) and the 2002 Hayman Fire (which charred another 138,000 acres) have prompted Denver Water to be more proactive in dealing with watershed health. The combination of these two fires, followed by significant rainstorms, resulted in more than 1 million cubic yards of sediment accumulating in Strontia Springs Reservoir, which is critically important to Denver Water's system as an intake point for the Foothills Treatment Plant – a facility that handles approximately 80% of Denver Water's supply.

Prior to the wildfires, Strontia Springs Reservoir contained approximately 250,000 cubic yards of sediment, which had been accumulating since 1983 when the dam was completed. Increased sediment creates operational challenges, causes water quality issues and clogs treatment plants.

Many parts of the U.S. are projected to experience increases in fire frequency and severity due to climate change. Therefore, to respond to the impacts of the Buffalo Creek and Hayman fires, and to mitigate impacts from future fires, Denver Water spent more than $26 million on water quality treatment, sediment and debris removal, reclamation techniques and infrastructure projects. In addition to impacts to the utility, the Hayman fire suppression costs for state and federal agencies were more than $42 million. The Hayman fire led to a loss of 600 structures, including 132 residences. Total insured private property losses were estimated at $38.7 million. Loss of wildlife habitat, aesthetics, tourism and recreation values are invaluable.

To mitigate future damage, Denver Water and the U.S. Forest Service Rocky Mountain Region are equally sharing an investment of $33 million over a multi-year period for restoration projects on more than 38,000 acres of National Forest lands (Denver Water 2014). The work will include mechanical thinning, fuel reduction, creating fire breaks, erosion control, decommissioning roads and reforestation.

 **United States Environmental Protection Agency**

CLIMATE READY
WATER UTILITIES
 ♨EPA

Group: SERVICE DEMAND & USE (DW/WW)

Return to Introduction

Changes in service demand associated with climate change may be driven or forced by altered volume and temperature of influent, as well as future challenges to meet the changing needs of agricultural and energy sectors. Clicking on either the drinking water or wastewater icon next to each impact will bring you to that particular Strategy Brief. Clicking on the Energy Management, Green Infrastructure or Water Demand Management icon will bring you to that Sustainability Brief.

Volume and Temperature Challenges

Climate change may lead to a growing imbalance in the demand for service and the ability of drinking water and wastewater utilities to meet it. Adaptation measures to identify additional water sources, improve efficiency of operations and promote conservation will provide benefits where changes in the supply and the scarcity of resources are of concern. Review the drinking water brief to learn more about how the Metropolitan Water District in Southern California has diversified its water supply to increase its reliability. Review the wastewater brief to learn more about how the City of Chicago implemented a green infrastructure program to manage stormwater runoff by reducing combined sewer overflows (CSOs).

Changes in Agricultural Water Demand

Changes in agricultural practices in response to climate change could significantly impact the ability of drinking water utilities to provide sufficient supply for their ratepayers. Rather than competing for limited resources during times of scarcity, these two sectors may have opportunities to collaborate on mutually beneficial solutions that meet their water needs. Review this brief to learn more about how Kern County, California uses water banking to increase its water storage capacity to offset impacts related to increased demand for water for agricultural purposes.

Changes in Energy Sector Needs and Energy Needs of Water Utilities

Changes in climate will impact both the energy sector directly as well as the energy needs of water utilities. The need for water used in energy generation is significant: thermoelectric power generation accounted for 49% of total water withdrawals in 2005 (USGS 2009). The energy required by the water sector for providing services is significant as well. Electricity accounts for about 75% of the cost of municipal water processing and transport and consumes about 4% of the nation's electricity (USGCRP 2009). Without cross-sectoral consideration of increased water and energy demands, future impacts from climate change may include higher operating costs, frequent loss of power and water shortages. Review this brief to learn more about how the Sonoma County Water Agency is reducing its greenhouse gas emissions to produce "carbon free" water, how Melbourne Water in Australia is supplying power utilities with reclaimed water for cooling and how the Albuquerque Bernalillo County Water Authority installed methane digesters to generate power for the utility.

Click to left of name to check off options for consideration; $'s (**$-$$$**) indicate relative costs
Click name of any option to review more information in the Glossary
⭐ **No Regrets options** - actions that would provide benefits to the utility under current climate conditions as well as any future changes in climate. For more information on No Regrets options, see Page 11 in the Introduction.
Click on the 💡, 🌱 or 🚰 icon to review the relevant Sustainability Brief.

ADAPTATION OPTIONS

✓	PLANNING	COST
⭐	Develop models to understand potential water quality changes (e.g., increased turbidity) and costs of resultant changes in treatment.	$
	Model sewer systems to understand the impact of higher groundwater infiltration on plant capacity and operating costs.	$
⭐	Use hydrologic models to project runoff and incorporate model results during water supply planning.	$
💡	Plan for alternative power supplies to support operations in case of loss of power.	$

Continued on page 2

INTRODUCTION

The Adaptation Strategies Guide for Water Utilities provides adaptation options for drinking water, wastewater and stormwater utilities based on region and projected climate impacts. While specific example adaptation options are described in the Guide, there is no one-size-fits-all solution for adaptation planning. Utilities will need to use the information included in this Guide to assist them in developing plans that contain adaptation options suited to their specific needs, taking into consideration their location, climate impacts of concern and available resources. Utilities should collaborate with state and federal authorities, interdependent sectors (i.e., energy, agriculture, forestry and wildlife) and other nearby utilities early in the adaptation planning process to ensure comprehensive and consistent planning.

Readers should use this Guide as an informational resource to identify potential strategies for adapting to climate change impacts. This Introduction provides background information to promote a better understanding of the adaptation planning process. The document then allows users to identify their relevant challenges by climate region and consider potential adaptation options to address these challenges. Finally, a _Worksheet for Adaptation Planning_ is provided to allow users to organize the information in this Guide and tailor it to their utility.

Why should a utility develop an adaptation plan?

Addressing climate change at a utility is a complex issue. Utilities are expected to encounter many challenges related to the impacts of projected future climate change. Because utilities also must address issues related to budget, aging infrastructure and other concerns, adaptation options that can address these issues in addition to providing greater resilience to climate impacts, are preferable. Climate change adaptation strategies may provide benefits such as more sustainable and efficient operations, cost savings, maintenance of adequate water supply and quality and the reduction of greenhouse gas emissions. Every utility has its own unique priorities and set of resources and will be impacted by climate change differently. Therefore, it is important to consider many different options and the range of benefits offered in order to develop a comprehensive adaptation plan that satisfies utility needs without overstretching resources.

Adaptation planning is not a new nor separate effort for managing utilities. Implementing adaptation strategies that provide multiple benefits can be integrated into current asset management, permit compliance, emergency response planning, capacity development and other decision-making processes at utilities.

Example of Approach to Incorporate Climate Change into Long-Term Planning:

Following deadly flash floods in 1997, the City of Fort Collins Utilities (FCU), Colorado, refocused its planning efforts around extreme precipitation events. FCU initiated a Climate Change Adaptation Study to examine possible future impacts of shifts in weather patterns. The purpose of this study was to understand the impacts of possible climate shifts and to design a framework to incorporate climate adaptation into FCU's ongoing asset management planning. As a mid-sized, combined drinking water, wastewater, stormwater and electric utility service provider, FCU has identified a need to adopt an integrated approach to adaptation and risk assessment. This approach would complement the integration of shifting weather patterns into utility design and management processes.

In summary, FCU has adopted the overarching goal of integrating adaptation planning into daily business practices by (1) embracing a dynamic, iterative process, (2) minimizing staff and resource burden by continually refining the process and (3) leveraging ties to asset management. One important step continued on next page

✓ PLANNING (continued)	COST
⊛ Conduct climate change impacts and adaptation training for personnel.	$
⊛ Develop energy management plans for key facilities.	$
⊛ Participate in community planning and regional collaborations related to climate change adaptation.	$-$$
⊛ Update drought contingency plans; include water conservation/water restriction requirements and actions for customers.	$
⊛⊛ Establish a relationship with the local power utility and work jointly on strategies to reduce seasonal or peak water and energy demands (e.g., water reclamation for use in power generation).	$
⊛⊛ Work with power companies to evaluate feasibility of using recycled water or alternative cooling methods to meet power plant needs.	$
⊛ Model agricultural water demand under future scenarios of climate change and projections of cropping types. Consider evaluating the use of recycled water for irrigation.	$-$$
Model or understand existing models of regional electricity demand under future scenarios of climate change and regional growth.	$

✓ OPERATIONAL STRATEGIES	COST
Conduct stress testing on wastewater treatment biological systems to assess tolerance to heat.	$$
⊛ Monitor current weather conditions, including precipitation and temperature.	$
⊛ Monitor surface water conditions, including streamflow and water quality in receiving bodies.	$
⊛ Finance and facilitate systems to recycle water, including use of greywater in homes and businesses.	$$-$$$
⊛ Improve energy efficiency of operations (e.g., installing more energy efficient pumps).	$$-$$$
⊛ Optimize operations by restricting some energy-intensive activities during the summer to times of reduced electricity demand (i.e., nighttime) and work with energy utility on off-peak pricing.	$$-$$$
Practice conjunctive use (i.e., optimal use of surface water and groundwater).	$$-$$$
⊛ Reduce agricultural and irrigation water demand by working with irrigators to install advanced equipment (e.g., drip or other micro-irrigation systems with weather-linked controls).	$$-$$$
⊛ Practice demand management through communication to public on water conservation actions.	$
⊛⊛⊛ Practice water conservation and demand management through water metering, leak detection and water loss monitoring, rebates for water conserving appliances/toilets and/or rainwater harvesting tanks.	$-$$

✓ CAPITAL/INFRASTRUCTURE STRATEGIES	COST
⊛ Acquire and manage ecosystems, such as forested watersheds, vegetation strips and wetlands, to buffer against floods and sediment and nutrient inflows into source waterways.	$$$
⊛ Build less energy-intensive treatment systems, such as using engineered wetlands.	$$$
⊛ Implement green infrastructure on site and in municipalities (e.g., green roofs, filter strips and more permeable building materials) to reduce runoff and associated pollutant loads into waterways.	$-$$$
⊛ Reduce inflow and infiltration into the sewer system by increasing control measures to decrease the volume of water to be pumped and treated.	$$-$$$
Build infrastructure needed for aquifer storage and recovery, either for seasonal storage or longer-term water banking, (e.g., recharge canals, recovery wells).	$$$
⊛ Diversify options to complement current water supply to include those that require less energy for treatment, conveyance, and distribution.	$$$
⊛ Expand current resources by developing regional water connections to allow for water trading in times of service disruption or shortage.	$$-$$$

✓	CAPITAL/INFRASTRUCTURE STRATEGIES	COST
⊛	Increase water storage capacity, including silt removal to expand capacity at existing reservoirs and construction of new reservoirs and/or dams.	\$\$-\$\$\$
⊚	Establish alternative power supplies, potentially through on-site generation, to support operations in case of loss of power.	\$-\$\$
⊛	Increase capacity for wastewater and stormwater collection, treatment and discharge, including redundancies to hedge against infrastructure losses and disruptions.	\$\$\$
⊛	Increase treatment capabilities to address water quality changes (e.g., increased turbidity).	\$\$\$
	Install effluent cooling systems (e.g., chillers, wetlands or trees for shading).	\$-\$\$
	Retrofit intakes to accommodate decreased source water flows or reservior levels.	\$\$-\$\$\$
	Build or expand infrastructure to support conjunctive use.	\$\$\$
⊚ ⊖	Build systems to reclaim wastewater for energy, industrial, agricultural or household use.	\$\$\$

 United States
Environmental Protection
Agency

CLIMATE READY
WATER UTILITIES
🌣EPA

VOLUME & TEMPERATURE CHALLENGES (DW)

Return to Introduction

Drought may increase in frequency and severity in some areas due to projected declining precipitation and increased loss of water from vegetation and evaporation. In areas dependent on snowmelt for water supplies, higher temperatures will reduce snowpack, thereby decreasing water storage. This combination results in decreased streamflow, reservoir safe yield and groundwater recharge. These impacts will reduce the available supplies for water systems dependent on surface water as well as groundwater, and potentially lead to service disruption. Diversifying water sources addresses some challenges associated with increased temperature, such as increased treatment costs associated with declining surface water quality. Groundwater often requires less treatment than surface water, and water recycling reduces the total amount of water that needs to be treated (Miller and Yates 2005).

CLIMATE INFORMATION

- By the end of the century, the average United States temperature is projected to increase by approximately 5°F to 10°F under the higher emissions scenario and by approximately 3°F to 5°F under the lower emissions scenario. The hottest day that occurs once every 20 years today is expected to occur once every two or three years over most of the U.S. (USGCRP 2014).

- Model projections of future precipitation indicate that southern areas, particularly the Southwest, will become drier (USGCRP 2014).

ADAPTATION OPTIONS

Click to left of name to check off options for consideration; $'s (**$-$$$**) indicate relative costs

Click name of any option to review more information in the Glossary

⭐ **No Regrets options** - actions that would provide benefits to the utility under current climate conditions as well as any future changes in climate. For more information on No Regrets options, see Page 11 in the Introduction.

Click on the 💡, 💧 or ☕ icon to review the relevant Sustainability Brief.

✔ PLANNING	COST
⭐ Develop models to understand potential water quality changes (e.g., increased turbidity) and costs of resultant changes in treatment.	$
⭐ Use hydrologic models to project runoff and incorporate model results during water supply planning.	$
⭐ Conduct climate change impacts and adaptation training for personnel.	$
💧 Participate in community planning and regional collaborations related to climate change adaptation.	$-$$
☕ Update drought contingency plans; include water conservation/water restriction requirements and actions for customers.	$

✔ OPERATIONAL STRATEGIES	COST
⭐ Monitor current weather conditions, including precipitation and temperature.	$
⭐ Monitor surface water conditions, including streamflow and water quality.	$
☕ Finance and facilitate systems to recycle water, including use of greywater in homes and businesses.	$$-$$$
Practice conjunctive use (i.e., optimal use of surface water and groundwater).	$$-$$$

✓ OPERATIONAL STRATEGIES (continued)	COST
● Reduce agricultural and irrigation water demand by working with irrigators to install advanced equipment (e.g., drip or other micro-irrigation systems with weather-linked controls). *(See example below)*	$$-$$$
● ● ● Practice water conservation and demand management through water metering, leak detection and water loss monitoring, rebates for water conserving appliances/toilets and/or rainwater harvesting tanks.	$-$$

✓ CAPITAL/INFRASTRUCTURE STRATEGIES	COST
● Acquire and manage ecosystems, such as forested watersheds, vegetation strips and wetlands, to regulate runoff.	$$$
Build infrastructure needed for aquifer storage and recovery (either for seasonal storage or longer-term water banking), e.g., recharge canals, recovery wells.	$$$
● ● Diversify options to complement current water supply, including recycled water, desalination, conjunctive use and stormwater capture. *(See example below)*	$$$
● Expand current resources by developing regional water connections to allow for water trading in times of service disruption or shortage. *(See example below)*	$$ $$$
● Increase water storage capacity, including silt removal to expand capacity at existing reservoirs and construction of new reservoirs and/or dams.	$$-$$$
● Increase treatment capabilities to address water quality changes (e.g., increased turbidity).	$$$
Retrofit intakes to accommodate decreased source water flows or reservoir levels.	$$-$$$
Build or expand infrastructure to support conjunctive use.	$$$
● ● Build systems to recycle wastewater for energy, industrial, agricultural or household use.	$$$

EXAMPLE

The Metropolitan Water District of Southern California (Metropolitan) is a wholesale water supplier for southern California. Metropolitan improved its Integrated Resource Plan in 1996 to enhance and diversify water supply reliability. Over the past decade, imported water supplies have been complemented by aggressive conservation programs, local water recycling, groundwater supplies, enhanced water storage and conveyance and water transfers. Metropolitan has helped develop more than 75 water recycling and groundwater recovery programs with local agencies through funding incentives. For example, the West Basin Municipal Water District receives secondary, treated wastewater from the City of Los Angeles, treats it to a tertiary level, and delivers it primarily for landscape irrigation and various industrial purposes. A portion of this water is injected to create an exclusion barrier against seawater intrusion into drinking water wells in the South Bay area. This project currently produces more than 20,000 acre-feet of water each year and is expected to expand to around 70,000 acre-feet of water each year by 2025. Moreover, Metropolitan has increased its storage capacity tenfold through the completion of both the Diamond Valley Lake in Hemet, new groundwater storage, and by acquiring contractual storage in state reservoirs. It has also been a leader in voluntary water transfers with agricultural districts. In August 2004, Metropolitan and the Palo Verde Irrigation District (PVID) executed a 35-year agreement under which individual landowners agree not to irrigate up to 29% of the valley's farm land, saving up to 111,000 acre-feet for other uses (Metropolitan 2010).

United States Environmental Protection Agency

CLIMATE READY
WATER UTILITIES
❧EPA

VOLUME & TEMPERATURE CHALLENGES (WW)

Return to Introduction

Climate models project that in the future, many areas are likely to receive less annual precipitation, but that when precipitation falls, it will be in fewer, more extreme rainfall events. These storm events wash sediment downstream and degrade water quality. Coupled with increases in algal growth resulting from the higher temperatures, generally diminished water quality in receiving waters may lead to more stringent requirements for wastewater discharges, higher treatment costs and the need for capital improvements. In some locations, lower flows and higher temperatures may impact ecosystems that are sensitive to temperature, requiring the utility to cool effluent prior to discharge.

CLIMATE INFORMATION

- The United States has recently seen unprecedented prolonged (multi-month) extreme heat events. The 2011 and 2012 extreme heat events set records for highest monthly average temperatures, hottest daytime maximum temperatures and warmest nighttime minimum temperatures (Karl et al. 2012). The likelihood of record-breaking temperature extremes is expected to increase in the future (USGCRP 2014).

- By the end of the century, the average U.S. temperature is projected to increase by approximately 5°F to 10°F under the higher emissions scenario and by approximately 3°F to 5°F under the lower emissions scenario. The hottest day that occurs once every 20 years today is expected to occur once every two or three years over most of the U.S. (USGCRP 2014).

- Model projections of future precipitation indicate that southern areas, particularly the Southwest, will become drier (USGCRP 2014). Lower volumes in surface water bodies may lead to higher pollutant concentrations.

ADAPTATION OPTIONS

Click to left of name to check off options for consideration; $'s ($-$$$) indicate relative costs
Click name of any option to review more information in the Glossary
⚝ **No Regrets options** - actions that would provide benefits to the utility under current climate conditions as well as any future changes in climate. For more information on No Regrets options, see Page 11 in the Introduction.
Click on the 🌐 or 💧 icon to review the relevant Sustainability Brief.

✓	PLANNING	COST
⚝	Develop models to understand potential water quality changes (e.g., increased turbidity) and costs of resultant changes in treatment.	$
	Model sewer systems to understand the impact of higher groundwater infiltration on plant capacity and operating costs.	$
⚝	Conduct climate change impacts and adaptation training for personnel.	$
💧	Participate in community planning and regional collaborations related to climate change adaptation.	$-$$

✓	OPERATIONAL STRATEGIES	COST
	Conduct stress testing on wastewater treatment biological systems to assess tolerance to heat.	$$
⚝	Monitor current weather conditions, including precipitation and temperature.	$
⚝	Monitor surface water conditions, including water quality in receiving bodies.	$

✓	CAPITAL/INFRASTRUCTURE STRATEGIES	COST
💧	Acquire and manage ecosystems, such as forested watersheds, vegetation strips and wetlands, to buffer against floods and sediment and nutrient inflows into source waterways.	$$$

Continued on page 2

✓	CAPITAL/INFRASTRUCTURE STRATEGIES (continued)	COST
🌱	Implement green infrastructure on site and in municipalities (e.g., green roofs, filter strips and more permeable building materials) to reduce runoff and associated pollutant loads into waterways. **(See example below)**	$-$$$
🌀	Reduce inflow and infiltration into the sewer system by increasing control measures to decrease the volume of water to be pumped and treated.	$$-$$$
⭐	Increase capacity for wastewater and stormwater collection, treatment and discharge, including redundancies to hedge against infrastructure losses and disruptions. **(See example below)**	$$$
⭐	Increase treatment capabilities and capacities to address more stringent treatment requirements (e.g., tertiary treatment).	$$$
	Install effluent cooling systems (e.g., chillers, wetlands or trees for shading).	$-$$

EXAMPLE

Like many cities that installed sewage collection systems prior to the 1930s, Chicago has a system that conveys both sewage and stormwater runoff. Large precipitation events can overwhelm the system, leading to combined sewer overflows (CSOs) that result in sewage flowing into the Chicago River, which degrades water quality in Lake Michigan. Chicago is building a deep tunnel system to expand capacity during flood events. This system will not be completed until 2019, and there are also concerns that extreme storm events will overwhelm even this expanded infrastructure. The city has therefore begun plans to implement a program to encourage the implementation of green infrastructure throughout the city, including:

- A Stormwater Management Ordinance mandates that as of 2008, any development that involves an area of 15,000 sq ft or creates a parking lot of 7,500 sq ft must retain the first half inch of rainfall on site or reduce the prior imperviousness by 15%.

- The Green Streets Program that has increased the proportion of the city shaded by tree canopy by 15%.

- The Green Roof Grant Program and Green Roof Improvement Fund that offers incentives for building green roofs. In 2007, the Chicago City Council allocated $500,000 to the Fund, and authorized the Department of Planning and Development to award grants of up to $100,000 to green roof projects within the City's Central Loop District.

- The Green Alley Program that began in 2006 and has started a series of pilot projects to test a variety of permeable paving materials to reduce flooding in alleys and increase infiltration of runoff. The City estimates that as of 2006, 1,900 miles of public alleys are paved with 3,500 acres of impervious cover.

These green infrastructure programs have been very successful. As of 2010, nearly 600,000 trees have been added to the cityscape and more than 4 million sq ft of green roofs have been installed on 300 buildings (U.S. EPA 2010). Green infrastructure can help both attenuate stormwater runoff and moderate the temperature of the water entering surface waters, and is thus an important climate change adaptation strategy.

 United States
Environmental Protection
Agency

CLIMATE READY
WATER UTILITIES
❄EPA

CHANGES IN AGRICULTURAL WATER DEMAND (DW)

Return to Introduction

Changing water needs of agricultural practices due to climate change could significantly impact the ability of drinking water utilities to provide sufficient supply. Competition could lead to shortfalls in water supply in the summer growing period, in particular. However, collaboration between the water and agricultural sectors can assist in meeting the water needs of both of these sectors. The following information describes strategies that water utilities can pursue while partnering with agricultural interests in their region.

CLIMATE INFORMATION

- By the end of the century, the average United States temperature is projected to increase by approximately 5°F to 10°F under the higher emissions scenario and by approximately 3°F to 5°F under the lower emissions scenario (USGCRP 2014).

- Increased temperatures mean increased evapotranspiration and crop water demands.

- By 2090, water demand is projected to increase by 42% under a lower emissions scenario and 82% under a higher emissions scenario, compared to 2005 levels. A large portion of the increase in demand is attributed to irrigation demands (Foti et al. 2012).

- Warming will increase the cultivation period of some crops (and irrigation water requirements); while others will have shorter periods due to heat stress (Backlund et al. 2008, USGCRP 2014).

ADAPTATION OPTIONS

Click to left of name to check off options for consideration; $'s (**$-$$$**) indicate relative costs

Click name of any option to review more information in the Glossary

⭐ **No Regrets options** - actions that would provide benefits to the utility under current climate conditions as well as any future changes in climate. For more information on No Regrets options, see Page 11 in the Introduction.

Click on the 💡, 💧 or ☕ icon to review the relevant Sustainability Brief.

✓	PLANNING	COST
⭐	Use hydrologic models to project runoff and incorporate model results during water supply planning.	$
⭐	Conduct climate change impacts and adaptation training for personnel.	$
💧	Participate in community planning and regional collaborations related to climate change adaptation.	$-$$
☕	Model agricultural water demand under future scenarios of climate change and projections of cropping types. Consider evaluating the use of recycled water for irrigation.	$-$$

✓	OPERATIONAL STRATEGIES	COST
⭐	Monitor current weather conditions, including precipitation and temperature.	$
☕	Finance and facilitate systems to recycle water, including use of greywater in homes and businesses.	$$-$$$
	Practice conjunctive use (i.e., optimal use of surface water and groundwater).	$$-$$$
☕	Reduce agricultural and irrigation water demand by working with irrigators to install advanced equipment (e.g., drip or other micro-irrigation systems with weather-linked controls).	$$-$$$
☕	Practice demand management through communication to public on water conservation actions.	$

 Continued on page 2

✓ OPERATIONAL STRATEGIES (continued)	COST
Practice water conservation and demand management through water metering, leak detection and water loss monitoring, rebates for water conserving appliances/toilets and/or rainwater harvesting tanks.	$-$$

✓ CAPITAL/INFRASTRUCTURE STRATEGIES	COST
Acquire and manage ecosystems, such as forested watersheds, vegetation strips and wetlands, to regulate runoff.	$$$
Build infrastructure needed for aquifer storage and recovery (either for seasonal storage or longer-term water banking), e.g., recharge canals, recovery wells. **(See example below)**	$$$
Diversify options to complement current water supply, including recycled water, desalination, conjunctive use and stormwater capture.	$$$
Expand current resources by developing regional water connections to allow for water trading in times of service disruption or shortage.	$$-$$$
Increase water storage capacity, including silt removal to expand capacity at existing reservoirs and construction of new reservoirs and/or dams.	$$-$$$
Build or expand infrastructure to support conjunctive use.	$$$

EXAMPLE

Water banking, a water leasing and trading tool used by the water sector to meet changing water demand, has been effectively applied in Kern County, California. Located at the southern end of the San Joaquin Valley, Kern County is one of the most productive agricultural counties in the nation, with more than 800,000 acres of irrigated farmland. The county is favorably situated for water banking in terms of geology, surface water supply, and delivery systems. Most of the water banks are highly permeable and well-suited for recharging underground aquifers. The earliest water banking programs began in the late 1970s and early 1980s with development of recharge ponds by the city of Bakersfield and the Kern County Water Agency. Today, the three major water banks have a combined storage capacity of about 3 million acre-feet – more than five times the amount of water in Millerton Lake, one of the larger reservoirs feeding the Central Valley surface-water system (Pacific Institute 2010).

United States
Environmental Protection
Agency

CLIMATE READY
WATER UTILITIES
⬥EPA

CHANGES IN ENERGY SECTOR NEEDS AND
ENERGY NEEDS OF UTILITIES (DW/WW)

Return to Introduction

Changes in climate will impact the energy sector directly and the energy needs of water utilities. Water usage in energy generation depends on many factors and is significant in scale. Thermoelectric power plants in Arizona, Colorado, New Mexico, Nevada, and Utah consumed an estimated 292 million gallons of water a day (MGD) in 2005, approximately equal to the water consumed by Denver, Phoenix, and Albuquerque, combined. The same year, thermoelectric power generation accounted for 49% of total water withdrawals in the United States – considerably more than the 31% withdrawn for agriculture (USGS 2009).

The energy required by the water sector to provide services is also significant. Electricity accounts for about 75% of the cost of municipal water processing and transport and consumes about 4% of the nation's electricity (USGCRP 2009). Surface water often requires more treatment than groundwater, and desalination is very energy intensive – energy accounts for 40% of the total desalination costs. Treated wastewater and recycled water (used primarily for agriculture and industry) require energy for treatment, but little for supply and conveyance (Cohen 2007, USGCRP 2009).

The following information describes strategies that water utilities can pursue while partnering with the energy sector in their region to reduce the amount of energy used in an effort to meet future water and energy needs. Without cross-sector consideration of increased water and energy demands, future impacts from climate change may include higher operating costs, frequent loss of power, and water shortages. These impacts will be most significant and likely during the summer, when water and electricity demand peak.

CLIMATE INFORMATION

- Summer electricity generation will likely be constrained by rising temperatures and water shortages. The efficiency of thermal power plants is sensitive to ambient air and water temperatures – higher temperatures reduce power outputs by decreasing the efficiency of cooling. Moreover, future water constraints on thermoelectric power plants are projected for Arizona, Utah, Texas, Louisiana, Georgia, Alabama, Florida, California, Oregon and Washington state by 2025 (USGCRP 2009).

- By 2030, water use for power production in the Rocky Mountain/Desert Southwest region is projected to grow by 200 MGD – an amount of water that could otherwise be used to meet the needs of 2.5 million people (WRA 2010).

- The total electricity demand for the U.S. is projected to increase 30% by 2035 compared to 2008 levels (US EIA 2010). Resultant increases in water demand will be a function of the fuel types used, cooling systems at thermoelectric plants and the rate at which existing plants are retired (DOE 2006).

- There will be disproportionately more electricity demand in the summer – it is projected that 4.5°F of warming would result in a 10% increase in net energy expenditures, while 9°F of warming would result in a 22% increase (Mansur et al. 2008).

- Higher water temperatures affect both the effectiveness of electric generation and cooling processes, and the ability to discharge heated water to streams from once-through cooled power systems due to regulatory requirements and concerns about ecosystem impacts (USGCRP 2014).

ADAPTATION STRATEGIES GUIDE FOR WATER UTILITIES

Continued on page 2

ADAPTATION
OPTIONS

Click to left of name to check off options for consideration; $'s ($-$$$) indicate relative costs
Click name of any option to review more information in the Glossary
⭐ **No Regrets options** - actions that would provide benefits to the utility under current climate conditions as well as any future changes in climate. For more information on No Regrets options, see Page 11 in the Introduction.
Click on the 💡, 💧 or 🖐 icon to review the relevant Sustainability Brief.

✓	PLANNING	COST
⭐	Use hydrologic models to project runoff and incorporate model results during water supply planning.	$
💡	Plan for alternative power supplies to support operations in case of loss of power.	$
⭐	Conduct climate change impacts and adaptation training for personnel.	$
💡	Develop energy management plans for key facilities.	$
🖐	Participate in community planning and regional collaborations related to climate change adaptation.	$-$$
💡 🖐	Establish a relationship with the local power utility and work jointly on strategies to reduce seasonal or peak water and energy demands (e.g., water reclamation for use in power generation).	$
💡	Work with power companies to evaluate feasibility of using recycled water or alternative cooling methods to meet power plant needs. *(See example 2 below)*	$
	Model or understand existing models of regional electricity demand under future scenarios of climate change and regional growth.	$

✓	OPERATIONAL STRATEGIES	COST
💡	Improve energy efficiency of operations (e.g., installing more energy efficient pumps). *(See example 1 below)*	$-$$$
💡	Optimize operations by restricting some energy-intensive activities during the summer to times of reduced electricity demand (i.e., nighttime) and work with energy utility on off-peak pricing.	$-$$$
	Practice conjunctive use (i.e., optimal use of surface water and groundwater).	$$-$$$
🖐	Practice demand management through communication to public on water conservation actions.	$
⭐ 💡 🖐	Practice water conservation and demand management through water metering, leak detection and water loss monitoring, rebates for water conserving appliances/toilets and/or rainwater harvesting tanks.	$-$$
💡 🖐	Practice water conservation and demand management to reduce energy demand and associated costs. *(See example 1 below)*	$-$$

✓	CAPITAL/INFRASTRUCTURE STRATEGIES	COST
🖐	Acquire and manage ecosystems, such as forested watersheds, vegetation strips and wetlands, to regulate runoff.	$$$
💡	Build less energy-intensive treatment systems, such as using engineered wetlands.	$$$
🖐	Reduce inflow and infiltration into the sewer system by increasing control measures to decrease the volume of water to be pumped and treated.	$$-$$$
	Build infrastructure needed for aquifer storage and recovery, either for seasonal storage or longer-term water banking, (e.g., recharge canals, recovery wells).	$$$
💡	Diversify options to complement current water supply to include those that require less energy for treatment, conveyance and distribution. *(See example 2 below)*	$$$

Example of Approach to Incorporate Climate Change into Long-Term Planning (continued)

towards this goal has been the use of the Joint Front Range Climate Vulnerability Study (CWCB 2011) as a source for climate scenarios that are based on model run results, including a range of possible futures: hot and dry; warm and wet; extreme drought; extreme precipitation and an average or "median conditions" scenario. From these scenarios, FCU has drawn information on impacts to water resources and potential flood events. For example, warmer and wetter winters may lead to decreased winter snowpack, increased rainfall and earlier spring melt and runoff for the area. With this information, FCU has (1) identified the risks related to these impacts, (2) considered consequences with respect to customers, operations and the environment through these risk assessments, and (3) evaluated adaptation options to address these risks and build a more resilient operation.

What is adaptation planning?

An integral part of increasing utility climate change resilience is to conduct a risk assessment and adopt an associated decision-support framework (**Figure 2.1**). This framework should be an iterative process of identifying projected impacts and challenges, assessing risks from these impacts, selecting and implementing adaptation options and then revisiting assessments when new information is available or when additional capacity to implement options is in place. The framework should also include other stressors besides climate change (e.g., changes in land use, population and regulatory changes).

Example of Adaptation Planning:

In 2012, Hurricane Sandy significantly challenged the operations of New York City's Department of Environmental Protection (NYC DEP), which provides drinking water, wastewater treatment and stormwater management services to over 9 million people. NYC DEP was able to continue to provide drinking water services throughout the storm, but 10 of the 14 wastewater treatment plants and 42 out of 96 pumping stations were damaged or lost power, resulting in the release of untreated or partially treated wastewater into local waterways. Hurricane Sandy was an example of the types of impacts that NYC DEP may continue to see in the future without adaptation.

For almost a decade, however, NYC DEP has taken a proactive approach to planning for climate change, beginning in 2004 when the utility partnered with Columbia University to conduct a climate vulnerability assessment and identify potential adaptation strategies, summarized in the 2008 Climate Change Program: Assessment and Action Plan. Based on the results of that assessment, the utility expanded their studies of climate impacts on both water supply, including the potential for increased turbidity and changes in streamflow and runoff related to reduced snow accumulation, and wastewater treatment, including the impacts of sea-level rise and storm surge on coastal infrastructure. Following Hurricane Sandy, NYC DEP expanded its wastewater study to conduct a more detailed, citywide risk assessment to identify which wastewater infrastructure is most vulnerable to flooding during extreme weather events both now and in the future.

According to this assessment, all of the city's 14 wastewater treatment plants have assets at risk to flooding from sea-level rise and storm surge impacts for the 2050s based on a "high end" sea-level rise projection of an increase of 30 inches, compared to a 2000-2004 baseline (NYC 2013). Fifty eight of the 96 pumping stations were also shown to be vulnerable. The City's potential exposure from the projected sea-level rise and related storm surge impacts was estimated to be $900 million at wastewater treatment plants and $220 million at pumping stations.

continued on next page

✓	CAPITAL/INFRASTRUCTURE STRATEGIES (continued)	COST
⊛	Expand current resources by developing regional water connections to allow for water trading in times of service disruption or shortage.	$$-$$$
⊛	Increase water storage capacity, including silt removal to expand capacity at existing reservoirs and construction of new reservoirs and/or dams.	$$-$$$
⚙	Establish alternative power supplies, potentially through on-site generation, to support operations in case of loss of power. *(See examples 1 and 3 below)*	$-$$
	Build or expand infrastructure to support conjunctive use.	$$$
⚙ ⊜	Build systems to reclaim wastewater for energy, industrial, agricultural or household use. *(See example 2 below)*	$$$

EXAMPLE 1

The Sonoma County Water Agency (SCWA) provides water to a population of over 600,000 in Sonoma and Marin counties in California. In 2006, SCWA realized that it was one of the largest energy users in Sonoma County, which drove the agency to take action to reduce energy usage and associated greenhouse gas emissions. As part of their response, SCWA established a goal to produce "carbon free" water by 2015 with respect to the electricity use for both water transmission and water treatment. To meet this goal, the agency is actively working to diversify its energy portfolio and reduce its energy and fuel needs through water conservation, system efficiency and procurement of renewable energy. Since 2006, SCWA has reduced water consumption by 28% through water conservation and water efficiency programs and has increased efficiency of water system operations by 18%. SCWA obtains power through a number of renewable green energy sources including: solar power (5% of energy use), hydropower (39% of energy use) and through a landfill gas-to-energy project (51% of energy use).

Through these initiatives, SCWA has reduced greenhouse gas emissions by 98% since 2006. The agency intends to achieve the final 2% reduction by switching meters to a local electricity provider, Sonoma Clean Power, which offers 100% renewable power at competitive rates. power supply. As a leader in climate action, SCWA has received multiple awards, including achieving platinum status with The Climate Registry, a non-partisan nonprofit organization that encourages organizations to voluntarily, accurately and consistently track their greenhouse gas emissions with a high level of accounting and integrity (SCWA 2012).

EXAMPLE 2

Melbourne Water (Victoria, Australia) is employing several strategies to expand water supply in response to climate change, given that precipitation and streamflow in its source areas may decline by 13% and 35%, respectively, by 2050. First, a major desalination plant is being constructed (due to be completed by the end of 2011), which will supply 150 billion liters of water – or about one third of the needed annual water supply – and will inherently be independent of hydrological variability. Second, the utility is upgrading its wastewater treatment plants to tertiary level, which will allow it to divert reclaimed water to power utilities that are currently using Latrobe Valley river water in power system cooling (Danilenko et al. 2010).

EXAMPLE 3

The Albuquerque Bernalillo County Water Utility Authority (ABCWUA) has installed methane digesters in its wastewater treatment plant, capturing methane and using it to generate both electricity and heat. In 2013, the treatment plant generated 21% of its power from utilizing waste methane instead of flaring it off as is commonly done (WRA 2010). Over 99% of all methane generated from this plant's sludge digestion process was beneficially used in 2013. An additional 6% of the total plant energy requirement was provided by renewable solar power.

SUSTAINABILITY BRIEF: GREEN INFRASTRUCTURE

Return to Introduction

Green infrastructure is an approach that uses either natural systems or engineered systems that mimic natural processes to control runoff and reduce water demand. The implementation of green infrastructure helps to address current challenges related to stormwater collection and treatment, limits floods from peak flows into surface waters following storms, augments groundwater supplies in shallow aquifers and supports source water protection efforts. Projected changes in precipitation patterns from climate change will exacerbate these existing challenges. Green infrastructure can be particularly effective when considered as part of a suite of options. For example, in conjunction with water conservation and infiltration/inflow reduction programs, it can be highly effective in reducing flow volumes.

BENEFITS OF GREEN INFRASTRUCTURE AS PART OF AN ADAPTATION PLAN

- **Increase collection capacity:** Reducing runoff volumes and rates through incorporation of green infrastructure within a service area decreases the overall flows into collection systems. This reduction of influent volumes can lower the frequency of combined sewer overflows and raw sewage backups as well as potentially reduce the need for infrastructure maintenance and expansion.

- **Increase resilience of service:** Facilitating groundwater recharge and reducing peak runoff flows may effectively reduce drought and flood-related service interruptions. Providing risk reduction through green infrastructure could improve the performance of other adaptation options to mitigate floods and droughts under projected climate conditions (e.g., combine rain gardens with stormwater storage to handle larger storms with current treatment capacity).

- **Enable incremental expansion of service:** Green infrastructure installations can be added as needed in areas that are not directly connected to existing infrastructure. In many cases, this decentralized approach allows for greater flexibility, faster implementation and lower costs than traditional grey infrastructure because it avoids building additional connections to the collection system.

- **Decrease carbon footprint:** Implementing green infrastructure projects can reduce collection and treatment needs, thereby reducing the utility's associated energy demand and greenhouse gas emissions. These projects can also help reduce the urban heat island effect—green roofs and other vegetation have been shown to keep buildings cooler during hot weather, reducing energy costs and emissions.

- **Leverage opportunities for co-benefits:** The costs of pursuing green infrastructure strategies may compare favorably to expanding or upgrading facilities when considering averted costs and additional benefits to the utility and larger community (e.g., air pollutant reductions and fewer odor complaints). The longer-term benefits from green infrastructure projects may become more apparent when costs and impacts of different options are assessed across multiple economic dimensions (e.g., public services, public health and ecosystem services).

- **Improve public image:** Many green infrastructure projects provide aesthetic enhancement to communities, particularly when compared with expansion of the built environment. Successful projects can make communities more attractive, increase property values, increase public safety and serve as visible reminders that a utility is pursuing adaptation holistically.

GETTING STARTED WITH GREEN INFRASTRUCTURE

Green infrastructure strategies can be pursued at various scales, either at utility facilities or throughout the community and service area. Depending on current or anticipated challenges and available resources, utilities may want to build on strategies that have been successful in the past or pursue new options. The following steps will help utilities get started with green infrastructure:

Continued on page 2

GETTING STARTED (continued)

- **Assess Current Challenges and Opportunities:** Assess the current state of green infrastructure in the utility and surrounding community and develop an awareness of local ordinances, regulations and building codes that may impact pursuit of green infrastructure. In addition, examine system performance in light of current and projected climate conditions (risk assessment) to identify potential needs with respect to runoff control and gauge the suitability of facilities or service area for green infrastructure implementation.

- **Evaluate Budgets and Funding Opportunities:** Weigh costs and benefits to consider green infrastructure as part of existing utility plans to improve and maintain facilities. Available funding for projects from government and other assistance programs may be a critical factor in identifying options and potential partners for both incremental improvements to current facilities and new projects (see funding section below).

- **Identify Strategies:** Based on the steps above, develop criteria and identify specific activities or projects to support pursuit of adaptation options from the tables below. Review available resources, including case studies of effectively implemented green infrastructure strategies at other utilities. Successful approaches will vary depending on location and utility size, but the experiences of similar utilities and local governments should help identify appropriate strategies.

- **Plan to Involve the Community:** Partner with community groups to implement and maintain green infrastructure projects more effectively. Potential partners for projects may include city government, local watershed groups, environmental nonprofits, local businesses, private developers, and community and neighborhood associations.

ADAPTATION OPTIONS (SUSTAINABLE PRACTICES)

Options for including Green Infrastructure in an overall adaptation strategy are provided in the tables below. Relative costs are provided on a qualitative scale ($ to $$$), and ⚙ indicates that an option could be considered a **"No Regrets"** strategy. For more information on No Regrets options, see Page 11 in the Introduction. Click on the ⚙ icon to review the relevant Sustainability Brief.

✓	GREEN INFRASTRUCTURE – PLANNING	COST
	Categorize existing and future conditions for land use and cover in watershed or collection area.	$
	Map land use with respect to impervious surfaces and potential sediment inputs.	$
⚙	Research use of green infrastructure to meet compliance needs, such as MS4 permits.	$
⚙	Compare green infrastructure approaches with alternatives in terms of costs and benefits in response to challenges.	$
⚙	Improve precipitation and collection models to inform runoff and influent predictions.	$
⚙	Design stormwater retention practices (e.g., rain gardens, green roofs) for flood prone areas (MS4s, CSO area) as part of planned improvements to collection in service area.	$
⚙	Conduct audit of facilities and overall system to determine suitability for new green infrastructure projects (e.g., green roof) and information needs to develop plans.	$
⚙	Train utility staff on green infrastructure technologies and maintenance.	$
⚙	Meet with community and local government officials to understand green infrastructure policies, practices and standards, assess local codes and regulations and to identify opportunities to overcome any existing barriers to green infrastructure.	$
⚙	Establish relationships with industries, local universities, businesses, and developers to collaborate on strategies to reduce stormwater runoff and flood damage (e.g., green roofs, bioswales) and provide necessary care-taking and maintenance.	$
⚙	Participate in community dialog to ensure green infrastructure is part of municipal modernization and upgrades to services and evaluate opportunities with other municipal services to leverage existing funds for green infrastructure opportunities.	$-$$
⚙	Investigate and define proper metrics, based on experiences of similar communities, to evaluate performance of green infrastructure projects.	$-$$

✓ GREEN INFRASTRUCTURE – OPERATIONAL	COST
⊛ Reduce infiltration / inflow by preventing illegal connections and leaks (e.g., grouting connections, sliplining, using watertight manhole covers) to reduce stormwater inflow volumes.	$-$$
Implement infiltration or recharge projects to reduce stormwater discharges or maintain groundwater table.	$$-$$$
⊛ Place filter strips or shoreline vegetation around surface water bodies or collection infrastructure vulnerable to sedimentation.	$$
⊛ Install rainwater harvesting and retention (e.g., rain barrels, green roofs) at current or planned facilities.	$$
Implement adaptive water rates to correspond with water supply.	$
⊛ Encourage water conservation on-site by employees and reduce water use on utility grounds by limiting irrigation and choosing native plants.	$
Develop communications package for customers promoting incentives and available equipment for rainwater collection and water conservation practices.	$-$$
⊛ Support green infrastructure development with large water consumers (e.g., industries, local universities) and land developers.	$$

✓ GREEN INFRASTRUCTURE – CAPITAL/INFRASTRUCTURE	COST
⊛ Improve stability and permeability of soil at facilities and in public areas to reduce runoff into stormwater collection system and surface water bodies.	$-$$
⊛ Build new stormwater retention structures (e.g., rain gardens, ponds) as part of planned improvements to collection in service area.	$$$
⊛ Support green infrastructure projects in the community as part of modernization and upgrades to services.	$$-$$$
⊛ Replace current paved surfaces (e.g., service roads and parking lots) with permeable surfaces.	$$-$$$

EXAMPLE 1

The City of Portland, Oregon, like many major U.S. cities, has experienced challenges related to combined sewer overflows and overall watershed health. The City Bureau of Environmental Services began a stormwater program in the 1990s and continues to implement green infrastructure solutions through the Grey to Green Program, which began in 2008, to help manage runoff and keep local rivers clean:

- From 1993 to 2011, the Downspout Disconnection program encouraged homeowners to redirect roof water to lawns and gardens, diverting approximately one billion gallons of stormwater per year from the combined sewer system.

- Green retrofits to streets have included landscaped curb extensions, swales, planted strips, pervious pavement and trees. Currently, the City has over 1200 green street facilities helping to manage stormwater.

- The City offers incentives via floor area bonuses to developers that pursue eco-roofs. As of July 2012, developers have built 355 eco-roofs that add more than 17.1 acres of green space.

- The Clean River Rewards program gives discounts on stormwater utility fees to homeowners that manage roof runoff by increasing pervious surfaces on their property.

- Active monitoring of projects has led to an increased understanding of effectiveness and the ability to refine projects as needed based on collected data.

- Acquisition of 318 acres of land and revegetation of natural areas are helping to control erosion and restore overall watershed health. Specific benefits of the current $5.6 million East Lents Floodplain Restoration project will include reduction of flooding frequency, improved wildlife habitat, stream bank stabilization, improved air quality and area

EXAMPLE 1 (continued)

revitalization. Phase 1 of the project is funded by a $2.7 million grant from the Federal Emergency Management Agency through its Pre-Disaster Mitigation Grant program

- Partnerships with local schools have allowed green projects to be used as educational opportunities for students and their families.

The success of green infrastructure in Portland is a result of many factors, including the City's multidisciplinary approach, the use of policy and incentives to encourage behaviors, and the use of monitoring data to continually refine approaches. The benefits of green infrastructure, including gains in watershed health, regulatory compliance, city livability and property values, are detailed in the 2010 report *Portland's Green Infrastructure: Quantifying the Health, Energy, and Community Livability Benefits*. Moving forward, the City of Portland plans to continue to use green infrastructure as an integral part of its stormwater management program.

EXAMPLE 2

The Milwaukee Metropolitan Sewerage District (MMSD) is complementing its past infrastructure investments with new investments in green infrastructure. These new projects have improved the effectiveness of the system during increasingly large storm events and maintained the provision of cost-effective water management services for the Milwaukee, Wisconsin region. Through prior efforts, MMSD decreased the number of system overflows from 50 to 60 per year before 1993 to a rate of just over two per year today. These efforts included 405 MG of inline storage constructed for just under $1 billion as part of the 1993 Water Pollution Abatement Program, 27 MG of additional storage built through the 27th Street Inline Storage System extension for $98 million and 80 MG of remote storage added for the northwest section of the service area for $161 million. The resulting $3 billion worth of infrastructure reduces the negative impact of wet weather events significantly. Looking forward, MMSD aims to eliminate overflows by 2035, in part, by integrating green infrastructure into overall system management and meeting its discharge permit condition to capture 1 million gallons per year over the next five years using green infrastructure. MMSD research results indicate that green infrastructure can be effective in reducing runoff and sediment loading while saving tunnel pumping costs, creating jobs, benefiting public health and improving community aesthetics. Recent MMSD accomplishments through green infrastructure practices include:

- Nearly 1 million gallons of stormwater prevented from entering the sewer system through distribution of almost 18,000 rain barrels throughout the city over the course of ten years. With the help of a local community service corps, MMSD purchases and retrofits reclaimed food grade barrels and then either sells or donates rain barrels to customers and community schools and groups.
- 8.4 acres of new green roofs in Milwaukee through a popular program, where MMSD matches funds to build green roofs on public and private buildings. A number of large manufacturing companies have constructed green roofs, and the public housing agency also considers green roofs on all new buildings.
- Protection of more than 2,400 acres of upstream wetlands and adjacent areas to filter stormwater pollution through the Greenseams® program, a partnership and land conservation program that aims to manage the percent of impervious surfaces in the watershed by keeping key open spaces open and undeveloped. MMSD spends an average of approximately $7,400 per acre on these plots of land, restores them if necessary, and then turns the land back over to the community or land trust to manage, ensuring that the land remains undeveloped.

More information can be found in MMSD's plan for *Sustainable Water Reclamation*.

EXAMPLE 3

New York City's Department of Environmental Protection (NYC DEP) is implementing a number of strategies to enhance the city's resilience to climate change, including the New York City Green Infrastructure Plan, a comprehensive 20-year effort to meet water quality standards. The goal of this plan is to capture stormwater runoff through green infrastructure projects (e.g., rooftop gardens, retrofitted buildings, swales) to reduce combined sewer overflow occurrences during heavy rain events. This cost-effective, community friendly program is an adaptive framework that can be modified to help New York City meet its adaptation goals. In March 2012, the plan was incorporated into a consent order with the State that would eliminate or defer $3.4 billion in traditional investments, and implementation of the initiatives in the plan would result

Continued on page 5

EXAMPLE 3 (continued)

in approximately 1.5 billion gallons of CSO reductions annually by 2030. The Green Infrastructure program leverages the established Greenstreets program, a collaboration between NYC Parks and NYC Department of Transportation that began in 1996. Greenstreets, which turns unused areas into green spaces, started as a way to beautify neighborhoods and improve air quality and, in 2010, became part of NYC DEP's Green Infrastructure Program to include additional stormwater benefits. Green infrastructure projects are currently being implemented and monitored in NYC DEP-designated priority sewersheds to understand and maximize benefits to the watershed.

In addition to green infrastructure, New York City is also expanding its Bluebelt program to enhance drainage in areas of the city that currently experience street flooding. These Bluebelts are natural areas that often enhance existing drainage corridors (such as streams, ponds and other wetland areas) to capture additional stormwater in place of installing new "grey" infrastructure. The first Bluebelt was constructed in Staten Island; almost 10,000 acres are currently in place. NYC DEP is currently constructing new Bluebelt systems in Staten Island and in Twin Pond, Queens, with plans to construct additional Bluebelts in Staten Island and the Bronx in the future (NYC 2013).

ADDITIONAL RESOURCES FOR GREEN INFRASTRUCTURE

PUBLICATIONS

- EPA's Green Infrastructure webpage
- Green Infrastructure (American Society of Landscape Architects)
- Sustainability in the Water Sector (International Water Association)
- Low Impact Development Center publications

- Stormwater Strategies (Natural Resources Defense Council)
- Rooftops to Rivers II (Natural Resources Defense Council)
- Integrating Valuation Methods to Recognize Green Infrastructure's Multiple Benefits (Center for Neighborhood Technology)

TOOLS

- EPA Green Infrastructure – Links to Modeling Tools
- Smart Growth/Smart Energy Toolkit: Low Impact Development (State of Massachusetts)
- National Low Impact Development Atlas (University of Connecticut)
- National Green Values™ Stormwater Management Calculator (Center for Neighborhood Technology)

- Green Roof Energy Calculator (Portland State University
- GIS Mapping Data on Existing Impervious Cover: National Land Cover Database
- International Stormwater BMP Database
- Stormwater Report website (Water Environment Federation)

FUNDING

- EPA Green Infrastructure: Funding Options
- EPA Clean Water State Revolving Fund (with information on EPA Green Project Reserve)
- EPA Drinking Water State Revolving Fund
- Financing for Environmental Compliance - Water Resources and Tools (EPA)

- EPA Tools for Financing Water Infrastructure
- USDA Rural Development Grants
- University of North Carolina Environmental Finance Center

 United States
Environmental Protection
Agency

CLIMATE READY
WATER UTILITIES
♺EPA

SUSTAINABILITY BRIEF: ENERGY MANAGEMENT

Return to Introduction

Significant amounts of energy are needed to support key water sector utility processes, including the pumping, production, conveyance, treatment, distribution, discharge and reuse of water. Energy management refers to strategies to reduce energy costs through changes in timing and amount of energy consumption while maintaining or improving services. Management strategies include reducing energy demand, improving water and energy efficiency of system operations, energy optimization and generating energy on-site via energy recovery methods or renewable sources. While many utilities pursue energy management as part of best practices in the industry, others are managing energy use in response to budget limitations and increased service demand. As climate continues to change, energy management will become increasingly integral to effective utility management, adaptation and mitigation.

BENEFITS OF ENERGY MANAGEMENT AS PART OF AN ADAPTATION PLAN

- **Improve efficiency and build energy independence:** Integrating more efficient equipment and processes into current operations, coupled with establishing independent energy supplies, reduces risks associated with service interruptions due to power outages and can improve predictability of future energy costs.
- **Increase operational flexibility and resilience of service:** Sustainable use of existing water and energy resources can reduce risks related to projected decreases in water supply and increases service demand. Energy management can increase utility efficiency and help balance overall energy and water needs, particularly as utilities begin to consider other adaptation options that may increase or decrease use.
- **Cost savings and opportunity to reinvest:** More efficient use of water and energy often generates a net savings which can be reinvested to help address other challenges such as the need for rate increases, the need to address gaps in funding or can be used to support additional adaptation efforts.
- **Decrease carbon footprint:** Implementing energy recovery or renewable energy projects can help increase sustainability by reducing demand on electric grids, use of electricity and associated greenhouse gas emissions.
- **Improve public image:** Communicating energy management practices to customers can establish a utility as a leader in pursuing financially and socially responsible actions. For example, wastewater utilities with successful on-site generation are viewed as resource recoverers as opposed to waste producers.

GETTING STARTED WITH ENERGY MANAGEMENT

Depending on the unique challenges and opportunities at each utility, energy management can be approached in several different ways. Utilities may want to build on strategies that have been successful in the past or pursue new options. The following steps, which echo the Plan-Do-Check-Act approach found in EPA's _Energy Management Guidebook for Wastewater and Water Utilities_, will help utilities get started with energy management.

- **Assess Current Energy Usage:** Conduct an energy assessment or energy audit to understand energy use and begin identifying areas for improvement. Use results of these assessments to examine existing utility goals and begin to refine these goals or establish new goals to improve energy efficiency (see Resources section below).
- **Evaluate Budgets and Funding Opportunities:** Determine availability of funds for energy management projects in short- and long-term budgets. If necessary, research and pursue funding available from state and federal assistance programs, foundations or community partners and energy performance contracting arrangements (see Resources section below).

Continued on page 2

GETTING STARTED (continued)

- **Identify Strategies:** Based on the steps above, develop criteria and identify specific activities or projects to support pursuit of adaptation options from the tables below. Review available resources, including case studies of effective energy management strategies at other utilities, to help identify appropriate strategies.

- **Plan to Involve the Community:** Reach out to customers to gauge interest in energy management and to discuss potential energy management and water conservation initiatives. If a number of options are available, community feedback may help in identifying demand management options that will have the greatest impact.

ADAPTATION OPTIONS (SUSTAINABLE PRACTICES)

Options for including Energy Management practices as part of an overall adaptation strategy are provided in the tables below. Successful approaches will vary depending on current energy sources and utility type. Relative costs are provided on a qualitative scale ($ to $$$), and ⊛ indicates that an option could be considered a "**No Regrets**" strategy. For more information on No Regrets options, see Page 11 in the Introduction.

Click on the ⊕ icon to review the relevant Sustainability Brief.

✓	ENERGY MANAGEMENT – PLANNING	COST
	Assess emissions footprint to develop a baseline for state or regional greenhouse gas emissions assessments and evaluations.	$-$$
⊛	Conduct an energy audit and set goals for energy use, conservation, recovery or alternative power supplies based on audit results.	$-$$
⊛	Develop an energy management team, including top-level management endorsement and support, and plan alternative power supplies to support operations in case of loss of power.	$
⊛	Assess energy implications (energy for treatment, conveyance and distribution) of any potential new source water (e.g., desalination plant, new wells).	$
⊛	Assess the marginal costs and payback periods for purchasing higher efficiency equipment as part of regular utility upgrades.	$
⊛	Develop and use hydrologic models to project runoff, understand potential water quality changes (e.g., increased turbidity) and costs of resultant changes in treatment and incorporate model results into water supply planning.	$
	Model energy demand or understand existing models of regional electricity demand under future scenarios of climate change and regional growth.	$
⊕	Estimate the reduction in greenhouse gas emissions resulting from water conservation and demand management.	$
⊕	Evaluate and compare the life cycle energy costs of potable and recycled water to gauge feasibility of systems to reclaim and reuse water, including use of greywater in homes and businesses.	$
⊛	Train personnel on energy efficiency and optimization practices and utility energy management goals and strategies, use short-term consumption forecasting and the effective use of automation.	$
⊛	Develop an energy management outreach plan and research opportunities for funding efficiency measures from state and local government assistance programs and other funding sources.	$
⊛	Establish relationship with local power utility and work jointly towards power purchase agreements (e.g., rate structure to encourage use during low-demand periods) and on strategies to reduce seasonal or peak water and energy demands (e.g., water reclamation for use in power generation).	$

✓	ENERGY MANAGEMENT – OPERATIONAL	COST
⊛	Monitor utility energy use and evaluate progress towards goals (water and cost savings, emissions reductions) and optimize operations by restricting some energy-intensive activities to times of reduced electricity demand (i.e., nighttime).	$-$$
⊛	Assess current energy use by identifying energy intensive processes (using sub-metering) and considering flows, load profiles, energy purchase design and operating schedules.	$-$$
⊛	Reduce wastewater treatment plant loading by using equalization basins and system-wide leak detection and repair to attenuate peak flows and loadings.	$-$$
⊛	Practice best building practices including installation of high-efficiency lighting, and maintenance of boilers, furnaces, high efficiency Heating Ventilation and Air Conditioning (HVAC) systems, motion sensor activating lighting, indirect fluorescent bulbs or using comprehensive lighting controls.	$-$$
⊛	Practice water conservation and demand management through public outreach, water metering and offering rebates for water conserving appliances and fixtures.	$-$$
⊛	Maintain vehicles to maximize fuel efficiency and reduce associated costs and emissions and purchase fuel efficient and alternative fuel vehicles when replacing older models.	$-$$
⊛	Install a Supervisory Control and Data Acquisition (SCADA) system for process monitoring and operational control (data for energy use optimization, detection of problems and compensation for seasonal or wet weather flows).	$$
⊛	Increase pumping efficiency by reducing and managing loads, modifying pumps, optimizing motor and drive selection, or pursuing automated control.	$-$$
⊛	Increase aeration efficiency by adding fine bubble aeration, improving surface aerators, installing more efficient motors, blower Variable Frequency Drives or automatic dissolved oxygen controls.	$$-$$$
⊛	Increase dewatering efficiency by replacing vacuum systems, installing premium motors or Variable Frequency Drives for water pumps.	$$-$$$
	Practice conjunctive use (i.e., optimal use of surface water and groundwater).	$$-$$$
⊛	Finance and facilitate water and wastewater projects to reclaim and reuse water, including use of greywater in homes, businesses and for irrigation needs (e.g., city parks).	$$-$$$

✓	ENERGY MANAGEMENT – CAPITAL/INFRASTRUCTURE	COST
⊛	Purchase energy efficient models when upgrading equipment (e.g., pumps, motors).	$-$$$
⊛	Establish alternative power supply via on-site power sources (renewable resources or energy recovery projects) or multiple grid supply lines.	$-$$
⊛	Build less energy-intensive treatment systems where feasible, including the use of natural systems such as engineered wetlands.	$$$
⊛	Implement green infrastructure on site to reduce energy use related to heating and cooling buildings.	$-$$$
⊛	Diversify options to complement current water supply to include those that require less energy for treatment, conveyance and distribution.	$$$
	Implement cogeneration technology to generate electricity and recover heat on site using methane off-gas from anaerobic digesters or from distribution (or collection) systems using turbines.	$$-$$$
	Build systems to reclaim and reuse wastewater for energy, industrial, agricultural or household use.	$$$

EXAMPLE 1

Sheboygan Regional Wastewater Treatment Plant is managing its energy use in a way that is beneficial to both the bottom line and the environment. In 2002, rising energy costs spurred this utility to conduct a study of its energy use, establish a baseline for current use and investigate opportunities to increase system efficiency and generate power. Over the next five years, a plan focused on these opportunities was developed and implemented, resulting in the following changes to the plant's system operations and efficiency:

- Motor upgrades and the installation of variable frequency devices reduced energy use by 157,000 KWh per year, resulting in an annual savings in energy costs of $5,300.
- A combined heat and power (CHP) system comprised of microturbines to generate 700 KW per year of electricity using methane gas that was previously flared off as waste. This generation meets 90% of the utility's overall energy needs, including provision of heat for the digesters and facility buildings.
- Through partnerships with the City of Sheboygan and local power utilities, Sheboygan Regional Wastewater Treatment Plant has been able to sell excess electricity generated by the CHP system back to the City.
- A $901,000 investment in upgrades to the aeration system included the purchase of two energy efficient blowers and air flow control valves for basins, which saves $63,000/year in energy costs.

Overall, the Sheboygan Regional Wastewater Treatment Plant has made a $2.5 million investment in these energy management projects with a 7- to 8-year payback period based on annual savings of about $500,000 in energy costs. By pursuing a more efficient energy system, Sheboygan Regional Wastewater Treatment is saving money and energy, reducing the release of waste products (methane gas) into the environment and has built important relationships with key partners in building sustainable practices across the City.

EXAMPLE 2

Waco Metropolitan Area Regional Sewerage System (WMARSS) serves approximately 175,000 people in the cities around Waco, Texas. The concept of sustainability at WMARSS involves reconsidering byproducts formerly considered "waste" as potential resources. The "waste to energy" initiative at WMARSS supports production of heat and energy from concentrated high-strength organics/fat, oil and grease (HSO/FOG). Using HSO/FOG increases the production of methane (or "biogas") from the anaerobic digestion process. Local food producers and restaurants provide 600,000 gallons of HSO/FOG per month to the WMARSS facility. This partnership also provides the additional benefit of keeping FOG out of sewer systems. The project expanded during the "Green Turkey Initiative" to accept fat, oil and grease from residents as well. From their $3.17 million investment, WMARSS now produces 600,000 cubic feet of biogas per day to supply one-third of the plant's electricity needs and 50% of the heating needs for its biosolids dryer/pelletizer.

WMARSS will continue to reap the benefits of its sustainable energy management practices through reduced costs, increased efficiency, improved customer service and reduced reliance on purchased electricity. WMARSS has also invested in other standard energy management practices, such as improvements to its aeration system at a cost of about $400,000 and a payback period of 2.4 years.

EXAMPLE 3

The Bureau of Environmental Services (BES) for the City of Portland, Oregon, has a history of improvements at its Columbia Boulevard Wastewater Treatment Plant that conserve and manage energy to reduce costs and demonstrate sustainable practices. Key areas of focus to manage energy have been to maximize the beneficial reuse of the digester gas (methane or biogas) produced in the anaerobic digestion process, improve energy efficiency through motor and lighting upgrades and control system improvements and prioritize energy efficiency improvements in new projects. A new building to house staff is also under construction and will be LEED Gold certified.

Currently, approximately 80% of the biogas produced in the anaerobic digesters is used to produce electricity and generate revenue. A large percentage of biogas is used in a combined heat and power (CHP) system which consists of two 850 KW reciprocating engines. This facility produces approximately 40% of the plant's energy demand and saves BES approximately $750,000 per year in energy costs. Biogas is also reused in boilers to provide heat in buildings and provide

Continued on page 5

Example of Adaptation Planning: *(continued)*

Based on the results of their risk assessment, NYC DEP developed a portfolio of strategies to protect wastewater assets from flooding impacts. Strategies include: dry flood-proofing buildings with watertight windows and doors, elevating equipment, making pumps submersible and protecting electrical equipment with watertight casings, constructing external flood barriers, installing sandbags temporarily, and providing backup power generation to pumping stations. Through strategically implementing a variety of strategies, considering utility resilience and return on investment, the utility could avoid 90% of risk citywide to wastewater treatment plants and ensure continuous service at pumping stations (NYC 2013).

1 Understand projected impacts and challenges	The first step in developing an adaptation plan is to gain a better understanding of how climate change, in combination with other stressors, may impact infrastructure and operations. These impacts could already be detectable or anticipated in the long term. The Climate Region Briefs included in this Guide provide an overview of national and regional climate projections from a recent assessment by the U.S. Global Change Research Program (2014) and list specific impacts relevant to water and wastewater utilities. See the references and links to additional information and resources.
2 Identify thresholds for failure or damage	The second step to consider is cataloging threshold conditions for critical assets, operational components and utility organization systems that may fail or suffer damage when challenged by climate change impacts. When compared to projected climate conditions, thresholds represent the capacity of the utility that may be bolstered through the implementation of adaptation plans. These thresholds can be determined through review of event and performance history, modeling of system performance or inspection of assets. For example, the elevation of coastal facilities, combined with precipitation totals could define the flood stage or combination of sea-level rise and storm intensity as thresholds for flooding.
3 Assess risks	After gaining a better understanding of both the thresholds for failure and the projected impacts, it is important to identify potential risks to infrastructure and operations. While there is no further guidance for conducting a risk assessment in this Guide, EPA's Climate Resilience Evaluation and Awareness Tool (CREAT) is available for free download and provides a framework for utilities to conduct a climate change risk assessment.
4 Determine adaptation options	Results from a risk assessment can be used to identify options that reduce system vulnerabilities. The Strategy Briefs included in this Guide provide general information on the impacts of climate change and lists of adaptation options that can be implemented to reduce potential consequences to operations and infrastructure. In addition to reducing risk, options should also be considered with respect to (1) current utility improvement plans and priorities and (2) current and projected available resources. For example, if assessments indicate high risk to coastal outfalls and pumps from flooding, then options to mitigate flood damage should be considered with respect to overall infrastructure planning and general system updates.
5 Implement and monitor	Following the design and implementation of any adaptation plan, utilities are encouraged to monitor conditions, compare results to projections and reassess both risk and adaptation options as new information becomes available. Monitoring should include remaining aware of new climate information and tools as they become available.

Figure 2.1. General process steps for adaptation planning. Steps are numbered based on the process described in this Guide. Other stressors (e.g., land use changes and population growth) contribute to the overall assessment and may, in turn, be altered by the adaptation options implemented. Steps "1 Understand projected impacts & challenges" and "4 Determine adaptation options" are addressed within this Guide.

EXAMPLE 3 (continued)

backup or supplemental heat to the cogeneration facility. Since the mid-1980s, approximately 25% of the biogas has been sold to a local industrial facility, generating $300,000 in annual revenue. The remaining biogas (20%) is currently flared as waste; however, BES is exploring plans to utilize this waste gas, include potential CHP expansion, production of a compressed natural gas for vehicles or selling this gas to a local natural gas utility.

BES has pursued multiple options for energy management, focusing on small projects that have both quick paybacks and continuing energy savings. In 2010, with the Energy Trust of Oregon's incentive program, BES completed two lighting retrofit projects that reduce energy costs by an estimated $10,000 annually with a payback period of less than one year. In 2011, BES upgraded the plant's compressed air system, which saves $17,000 per year. A current project will optimize dissolved oxygen (DO) control in the activated sludge process, supporting more precise DO set points and saving approximately $30,000 annually. The payback period for these improvements, after incentives, is estimated to be seven years.

ADDITIONAL RESOURCES FOR ENERGY MANAGEMENT

PUBLICATIONS

- Evaluation of Energy Conservation Measures for Wastewater Treatment Facilities (EPA)
- Ensuring a Sustainable Future: An Energy Management Guidebook for Wastewater and Water Utilities (EPA)
- Energy Efficiency for Water Utilities (EPA)
- Water Reuse Guidelines (EPA)
- EPA's Energy Efficiency for Water and Wastewater Utilities website

- EPA's State and Local Climate and Energy website
- EPA's WaterSense website
- Evaluation of Combined Heat and Power Technologies for Wastewater Facilities (Columbus Water Works)
- Water Consumption Forecasting to Improve Energy Efficiency of Pumping Operations (Water Research Foundation/California Energy Commission)

TOOLS

- EPA Portfolio Manager (ENERGY STAR)
- EPA Energy Use Assessment Tool
- Water Energy Sustainability Tool (UC Berkley)

- Water Conservation Tracking Tool (Alliance for Water Efficiency)
- Water Energy Simulator (Pacific Institute)

FUNDING

- Energy Efficiency RFP Guidance for Water-Wastewater Projects (Consortium for Energy Efficiency)
- EPA Clean Water State Revolving Fund (with information on EPA Green Project Reserve)
- EPA Drinking Water State Revolving Fund

- USDA Rural Development Grants
- Financing for Environmental Compliance – Water Resources and Tools (EPA)
- EPA's Tools for Financing Water Infrastructure

 United States
Environmental Protection
Agency

CLIMATE READY
WATER UTILITIES
♻EPA

SUSTAINABILITY BRIEF: WATER DEMAND MANAGEMENT Return to Introduction

Population growth as well as climate change impacts, such as increasing temperatures and the increased risk of prolonged periods of drought, can contribute to unsustainable demands on water services and thus increase the risk of water shortages. Water demand management encompasses both water efficiency and conservation practices and can occur on the supply side (related to drinking water utility actions to increase the efficiency of delivering water to customers) and the demand side (related to customer actions to reduce the amount of water used in homes and businesses). Water conservation is a cost-effective method that can help meet current and future water needs in a sustainable manner. While many utilities pursue water demand management as a part of best practices in the industry, others are looking to these practices in response to budget limitations and increased service demand. As climate continues to change, water efficiency and conservation will become increasingly integral to effective utility management, adaptation and mitigation.

BENEFITS OF WATER DEMAND MANAGEMENT AS PART OF AN ADAPTATION PLAN

- **Increase operational flexibility and resilience of service:** Sustainable use of existing water resources can reduce risks related to projected decreases in water supply and increases in service demand. Water conservation reduces the need to develop new source water supplies or to expand the infrastructure at water and wastewater facilities. Water efficiency and conservation programs can preserve natural resources and increase the sustainability of water supplies, leaving more water for future use and improving the ambient water quality and aquatic habitat.

- **Cost savings and opportunity to reinvest:** More efficient use of water often reduces operating and treatment costs, resulting in a net savings which can be reinvested to help address other challenges – such as the need for rate increases, the need to address gaps in funding or can be used to support additional adaptation efforts. When faced with potential water shortages, developing and implementing water efficiency and conservation measures almost always involves a lower cost than developing a new water source or expanding water or wastewater infrastructure to meet demand or other goals.

- **Deferred and avoided capital investments:** Water demand management practices will often allow the utility to continue to meet water demand without needing to expand existing facilities or build new facilities. Water demand management can also extend the life of existing facilities.

- **Maintain environmental benefits of water resources:** Reduced water consumption helps to maintain reservoir water levels and groundwater tables, and supports the use of lakes, rivers and streams for recreation and wildlife. When use of these resources reduces surface or groundwater levels, natural and human pollutant levels can increase and threaten human and ecological health. Using water more efficiently helps maintain supplies at safe levels, protecting human health and the environment.

- **Decrease carbon footprint:** The delivery of water requires energy to pump, treat and distribute water. End users also use energy to heat water for certain uses. Implementing water efficiency and conservation projects can reduce the amount of water withdrawals from sources and demand on wastewater services, thereby reducing energy needs and associated greenhouse gas emissions. Use of more water efficient products by customers can also decrease energy needed to heat water.

- **Improve public image:** Communicating utility actions to increase water efficiency and encouraging water conservation practices to customers can establish a utility as a steward of local water resources and a leader in pursuing financially and socially responsible actions.

GETTING STARTED WITH WATER DEMAND MANAGEMENT

Depending on the unique challenges and opportunities at each utility, water efficiency and conservation can be approached in several different ways. Utilities may want to build upon strategies that have been successful in the past

    Continued on page 2

GETTING STARTED (continued)

or pursue new options. The following steps will help utilities get started. More information on developing a water conservation plan for your utility can be found in EPA's Water Conservation Plan Guidelines.

- **Assess Current Water Usage:** Review overall current and historical water use by customer class and consider how usage may need to be reduced considering changes in the future availability of water. Conduct a water audit using the International Water Association (IWA)/American Water Works Association (AWWA) Water Audit Method to understand water use and non-revenue water (which includes real and apparent water losses), and begin identifying areas for improvement. Use results of these assessments to examine existing utility goals and begin to refine these goals or establish new goals to improve water efficiency and customer conservation practices (see Resources section below).

- **Evaluate Budgets and Funding Opportunities:** Determine availability of funds for water efficiency and conservation projects in short- and long-term budgets. If necessary, research and pursue funding available from state and federal assistance programs, foundations or community partners and energy performance contracting arrangements (see Resources section below).

- **Identify Strategies:** Based on the steps above, develop criteria and identify specific activities or projects to support pursuit of adaptation options from the tables below. Review available resources, including case studies of effective water conservation strategies at other utilities, to help identify appropriate strategies.

- **Plan to Involve the Community:** Water conservation offers many benefits to customers and society. Involving the community in goal development and implementation serves an important educational function, and can enhance the success of the program. Educational programs for utility employees, customers and school children are vital to the success of a water conservation program.

- **Become a WaterSense partner:** EPA's WaterSense program provides excellent resources on water conservation practices, as well as communication tools.

ADAPTATION OPTIONS
(SUSTAINABLE PRACTICES)

Options for including Water Demand Management in an overall adaptation strategy are provided in the tables below. Relative costs are provided on a qualitative scale ($ to $$$), and ⊛ indicates that an option could be considered a **"No Regrets"** strategy. For more information on No Regrets options, see Page 11 in the Introduction.

✔	WATER DEMAND MANAGEMENT – PLANNING	COST
	Identify a water efficiency coordinator for the utility and establish a water efficiency team, including top-level management endorsement and support.	$
	⊛ Conduct training for utility staff on water conservation policies and goals.	$
	⊛ Review water use by customer classes, set goals for use and develop a water conservation plan. Consider strategies for reducing water use by municipal, residential, commercial, institutional and industrial users.	$
	Review the utility rate structure and ensure that it encourages water efficiency. Consider using non-promotional pricing structures such as inclining tier rates, excess use rates and seasonal rates.	$-$$
	⊛ Conduct water loss audits using the IWA/AWWA Water Audit Method (see resources) and set goals for reductions in water loss that will inform a water loss management program. Water losses may be real (e.g., from leaks) or apparent (e.g., meter inaccuracy, unauthorized consumption).	$-$$
	⊛ Conduct an audit of water use in utility operations (e.g., solids handling in wastewater treatment operations) and determine opportunities to reduce water use.	$
	Conduct water use audits of homes, businesses and industries to provide information to customers about how water is used and how usage can be reduced.	$$
	Incorporate water conservation actions and associated cost savings into utility water demand forecasting and integrated resources planning.	$
	Estimate the reduction in greenhouse gas emissions resulting from water conservation and demand management measures.	$

Continued on page 3

✓ WATER DEMAND MANAGEMENT – PLANNING (continued)	COST
Evaluate and compare the life cycle energy costs of potable and recycled water to gauge feasibility of a system to reclaim and reuse water, including use of greywater in homes and businesses. Consider evaluating the use of recycled water for irrigation.	$-$$
Develop a comprehensive outreach and education program for residential, industrial and commercial customers as well as schools to encourage water conservation. Provide water conservation literature to new customers when they apply for service.	$
Include water conservation measures in the utility's drought contingency plan and establish conditions where it may be necessary to implement emergency water conservation measures and water restrictions throughout the service area. Determine how to communicate acceptable and unacceptable water usage to customers.	$-$$
Research opportunities for financing water efficiency measures from state and local government assistance programs and other funding sources.	$
Establish a process to monitor utility water use and evaluate progress towards goals (utility water loss reductions and customer water usage, cost savings, energy savings and emissions reductions).	$-$$

✓ WATER DEMAND MANAGEMENT – OPERATIONAL	COST
Implement a universal metering program, including plans for meter testing, repair and periodic replacement. Install water meters in previously unmetered areas (if rate structure is based on metered use).	$$
Implement a water-loss management program for leak detection and repair as well as water loss for the water transmission, delivery and distribution system.	$-$$
Minimize the water used in space cooling equipment in accordance with manufacturers' recommendations. Shut off cooling units when not needed.	$$
Ensure that fire hydrants are tamper proof to eliminate unauthorized consumption of water.	$$
Provide retrofit kits for residences and businesses for free or at cost. Kits may contain WaterSense labeled faucet aerators, showerheads, leak detection tablets and replacement valves. Consider offering an installation program to retrofit plumbing devices in the service area.	$-$$
Offer incentive programs (rebates/tax credits) to homeowners and businesses to encourage replacement of plumbing fixtures and appliances with water-efficient models.	$$
Promote water-efficient landscape practices for homeowners and businesses, especially those with large, irrigated properties. Practices include use of native plants, landscape renovation to reduce water use, use of irrigation professionals certified by a WaterSense labeled program and more efficient irrigation.	$-$$
Reduce agricultural and irrigation water demand by working with irrigators to install advanced equipment (e.g., drip or other micro-irrigation systems with water-linked controls).	$$-$$$
Finance and facilitate water and wastewater projects to reclaim and reuse water, including use of greywater in homes, businesses and for irrigation needs (e.g., city parks).	$$-$$$
If permitted by local ordinances, encourage industrial and commercial customers to harvest rainwater and to collect condensate from large cooling systems to be used on-site for irrigation and other non-potable uses.	$-$$
Enforce or support regulations, ordinances or terms of service that prohibit water waste and address irrigation and other design inefficiencies and misuses of water.	$-$$

 Continued on page 4

✓	WATER DEMAND MANAGEMENT – CAPITAL/INFRASTRUCTURE	COST
	Install high-efficiency WaterSense labeled faucet aerators, showerheads and labeled toilets, or retrofit water-saving devices in municipal buildings. Replace municipal appliances or equipment with water-saving models at the end of their life cycle.	$
	Eliminate "once-through" cooling of equipment with municipal water by recycling water flow to a cooling tower or by utilizing air-cooled equipment.	$$-$$$
⊛	Reduce inflow and infiltration into the sewer system by increasing control measures to decrease the volume of water to be pumped and treated.	$$-$$$
	Replace or rehabilitate finished water storage tanks to minimize water loss.	$$-$$$
	Replace broken water meters or upgrade existing meters with automatic meter reading systems, advanced metering infrastructure, smart meters and meters with built-in leak detection. Install backflow prevention devices in conjunction with meter replacement.	$$-$$$
	Build systems to reclaim and reuse wastewater for energy, industrial, agricultural or household use.	$$$

EXAMPLE 1

The NYC Department of Environmental Protection (NYCDEP) has implemented a Water Demand Management Plan with the goal of reducing water consumption by 5% by 2020. The plan has five key strategies, one of which focuses on improving water efficiency at public facilities, including NYCDEP's 14 wastewater treatment plants. In 2012, the City designed an audit to review water usage at the wastewater plants, which consume approximately 7.3 million gallons per day (MGD) of potable water. The audits found that water use as a percent of dry weather flows varied from 0.21 to 4.77%. In addition to identifying opportunities to reduce water in different plant processes, the audits also evaluated the potential for replacing city potable water with effluent water for some processes. Overall, NYCDEP estimates that 2.1 MGD could be saved through efficiency actions. Another important outcome from the audit indicated that between 4 to 68% of each plant's pump process water usage could be decreased by replacing old pumps that use packing with pumps that use mechanical seals. In response, NYCDEP has changed its design guidelines to specific use of mechanical seals for any new pumps.

For more information, visit: http://www.nyc.gov/html/dep/html/ways_to_save_water/index.shtml.

EXAMPLE 2

The San Antonio Water System (SAWS) in Texas has an extensive and widely respected water conservation program. SAWS has fully integrated water conservation into its water resource management planning and factors potential savings into water and wastewater capital improvement plans. In large part due to prudent and strategic conservation programs starting in the 1980s, SAWS has eliminated the demand for a water supply project of approximately 120,000 acre-feet per year. This proactive planning approach has saved the utility billions of dollars in capital expenses and annual operating expenses. As SAWS plans to meet the demands of an increasing population, the utility is projecting that water conservation and recycling in a drought year will account for 20% of its water portfolio by the year 2030. Reasonable drought restrictions on discretionary uses play an important role in the utility's drought management plan, when mandatory measures are imposed to ensure reliable water service for essential uses such as indoor consumption, as well as commercial and industrial applications.

For more information, visit: http://www.saws.org/conservation/.
http://www.saws.org/Your_Water/WaterResources/2012_WMP/.

EXAMPLE 3

The Southern Nevada Water Authority (SNWA) and its member agencies supply water to approximately 2 million people and more than 40 million annual visitors. The SNWA currently draws about 90 percent of the community's water supply from the Colorado River via Lake Mead, the largest man-made reservoir in the United States. During the past decade, the impact of drought has caused Lake Mead elevations to decline by more than 100 feet, representing a storage loss of more than 4 trillion gallons. The SNWA is concerned about further reductions in streamflow from climate change and increased demands.

 Continued on page 5

EXAMPLE 3 (continued)

A study conducted by the Bureau of Reclamation and the Colorado River Basin States projected a 3.2 million acre-foot annual imbalance between supply and demand for the Colorado River Basin by 2060 (Bureau of Reclamation 2012). For the SNWA, reduced Colorado River streamflow would result in lower levels in Lake Mead, the potential loss of the ability to withdraw water from existing intakes, reduced water quality at withdrawal locations, and increased power requirements to pump water a greater vertical distance..

To ensure a reliable supply for residents and visitors into the future, the SNWA more than a decade ago launched an extensive water conservation program; it is also investing in significant infrastructure enhancements related to its Lake Mead intakes. Demand management practices (i.e., education, incentives, regulation and rates) have reduced consumptive water use by 32 percent since 2000, even as the population has increased by nearly half a million. Examples of successful strategies include: incentives for homeowners and commercial properties that convert turf to water efficient landscapes; working with landscapers in the area to provide them with water-efficient irrigation technology; rebates on pool covers; and time/day restrictions on landscape irrigation, including for commercial customers. Infrastructure enhancements include the completion of a second intake, ongoing construction of a third deeper intake in Lake Mead, additional water treatment capacity using ozone, and distribution system expansions.

The SNWA is also using EPA's Climate Resilience Evaluation and Awareness Tool (CREAT) to evaluate a number of physical adaptation measures to address the impacts related to declining lake levels. These options include infrastructure improvements to ensure operability at lower lake levels and constructing a new intake that can withdraw water from deeper in the lake where the water is cooler and of higher quality.

Despite the agency's successful conservation strategies, reservoir levels may continue to decline. Even if SNWA stopped withdrawing water entirely for a year, Lake Mead would only rise by approximately 3 feet, given its current elevation, which would do little to offset the 100 feet of decline that has been seen in the past 14 years. Recognizing this, SNWA has made it a priority to work with the other states that rely on the Colorado River and Mexico to develop innovative solutions through key partnerships (Bureau of Reclamation 2012).

For more information, visit: http://www.snwa.com/ws/resource_plan.html.

ADDITIONAL RESOURCES FOR WATER CONSERVATION

PUBLICATIONS

- EPA's WaterSense website*
- Water Audits and Water Loss Control for Public Water Systems (EPA)
- Water Conservation Communications Guide (American Water Works Association [AWWA])
- Water Conservation Resource Community** (AWWA)
- Water Loss Control Resource Community** (AWWA)
- Water Sense Guidelines for Preparing Water Conservation Plans (EPA)

- Control and Mitigation of Drinking Water Losses in Distribution Systems (EPA)
- Texas Water Development Board Water Conservation Best Management Practices for Municipal Water Providers
- Colorado Waterwise Guidebook of Best Practices for Municipal Water Conservation
- Water Efficiency Resource Library (Alliance for Water Efficiency)

*Utilities that sign up as a WaterSense promotional partner (free) can receive access to tools to help them carry out water conservation programs.

**Requires non-AWWA members to register (free) to access some information.

CUSTOMER OUTREACH EXAMPLES

- Save Dallas Water and Save Tarrant Water – Regional efforts of North Texas Utilities
- Save Our Water – California Department of Water Resources and the Association of California Water Agencies

- Be Water Wise – Metropolitan Water District of Southern California and Southern California Water Agencies
- Saving Water Partnership – Seattle and King County utilities

TOOLS

- AWWA Water Audit Software
- Water Conservation Toolbox for Water Suppliers (Metropolitan Council, Minnesota)
- Water Conservation Tracking Tool (Alliance for Water Efficiency)

FUNDING*

- EPA Clean Water State Revolving Fund (with information on EPA Green Project Reserve)
- EPA Drinking Water State Revolving Fund
- EPA's Tools for Financing Water Infrastructure
- Utility Financial Sustainability and Rates Dashboard (University of North Carolina Environmental Finance Center)
- USDA Rural Development Grants
- EPA's Catalog of Federal Funding Sources for Watershed Protection
- Bureau of Reclamation Water & Energy Efficiency Grants (for Reclamation states)

*Many states have funding programs that can fund water efficiency and conservation activities.

Adaptation planning involves more than just a review of options for facility owners and operators to consider. Several technical and informational resources are required to support planning. For example, inundation maps, precipitation projections and flood models may all need to be employed in the determination of thresholds for flooding and the assessment of adaptation options to mitigate losses. Utilities can access this information through a number of resources (see *Example Resources*).

How does a utility identify adaptation strategies for consideration?

Historically, utilities have applied the assumption that, while observed temperature and precipitation conditions may exhibit large variations, the variability and average conditions will remain consistent into the future. This assumption, often referred to as stationarity, will be compromised as climate changes. Many climate models project that future climate conditions (e.g., intensity of precipitation events, sea-level rise, temperature increases) may experience increased variability compared to that seen in the past; historical data and trends may no longer be accurate indicators for future climate conditions. Utilities should therefore adopt a flexible and iterative approach when considering what adaptation options to implement, and ensure that strategies are complementary to capacity building, emergency response activities, capital planning and sustainability planning. An adaptive approach will result in robust decision-making that builds operations that are successful regardless of the climate impacts faced by a utility.

In addition to the shift from stationarity in climate, utilities are increasingly recognizing that future energy prices and ecological conditions may not be predictable based on historical observations. These shifts may require utilities to change how they operate and manage their resources to ensure that they can withstand and adjust to these changes. Sustainability at a water utility is maintained through practices that address today's needs while ensuring continued and long-term provision of clean and safe water. Many sustainable practices offer opportunities to address climate-related challenges in a socially, economically and environmentally responsible way. The Sustainability Briefs included in this Guide address areas of overlap between adaptation and three types of sustainable practices: energy management, green infrastructure and water demand management.

Energy management can reduce operational costs, reduce greenhouse gas emissions and increase service flexibility. Practices that are generally considered part of energy management include any action taken to reduce, optimize or increase energy efficiency, including service demand reduction, or generation of energy on-site. These practices also provide opportunities to engage stakeholders in the long-term, forming collaborative partnerships that support a more sustainable community. From an adaptation planning perspective, utilities should consider the energy use implications of new decisions or operational changes and consider how alternate or additional options may limit any increase in energy needs. To access the Energy Management Sustainability Brief, click on this icon.

Another sustainable strategy, ***green infrastructure***, involves the use of natural systems (or engineered systems that mimic them) to help control runoff, capture stormwater and reduce water demand. Some common green infrastructure practices include green roofs, rain gardens, land acquisition and using permeable pavements. As an adaptation strategy, green infrastructure can help address both current and projected challenges related to stormwater management while also complementing the use of more traditional, grey infrastructure. One key advantage of green infrastructure is that projects can be phased in gradually, allowing projects to be adjusted as necessary. Pursuit of green infrastructure activities also promotes collaboration with stakeholders and local governments when considering the benefits of community-scale implementation. To access the Green Infrastructure Sustainability Brief, click on this icon.

Water demand management practices can increase the sustainability and long-term availability of water supplies. Climate impacts, such as increasing temperatures and the increased risk of prolonged periods of drought, combined with non-climate impacts such as population growth, can contribute to unsustainable demands on water services, increasing the risk of water shortages. Water demand management encompasses both water efficiency and conservation practices and can occur both on the supply side (related to drinking water utility actions to increase the efficiency of delivering water to customers) and the demand side (related to customer actions to reduce the amount of water used in homes and businesses). As a part of an adaptation plan, water demand management practices can allow a utility to continue to meet demand for services while reducing the need to develop new source water to expand existing supplies. To access the Water Demand Management Sustainability Brief, click on this icon.

Example of Sustainability Planning

Camden County Municipal Utilities Authority (CCMUA) in Camden, New Jersey is responding to rising energy costs, climate change and population growth by examining and improving system efficiency through a number of sustainable or "green" initiatives. To ensure the long-term viability of their operations, CCMUA has set four goals: (1) optimize water quality, (2) improve air quality, (3) minimize costs and (4) reduce energy and carbon consumption. It was a priority for CCMUA to not increase rates for its customers, so all of these initiatives and planned activities listed below are considered rate neutral, or in the case of some of the energy efficiency initiatives, actually allowed for lower rates.

CCMUA plans to switch to 100% green energy sources by 2017. The first step towards this goal was to minimize energy use in its system by reducing infiltration/inflow, using gravity connections to replace municipal pumping stations, implementing electric peak shaving, using heating loops and energy-efficient equipment and lighting and installing catalytic converters to improve air quality by reducing emissions. These efforts to minimize energy use at the utility were completed using a $10 million low interest State Revolving Fund (SRF) loan. The utility sees about $600,000 in energy savings per year, which is greater than the yearly payments to repay the SRF loan. These upgrades were done at strategic times when equipment already needed to be updated or replaced.

To get closer to its goal of 100% green energy, CCMUA installed a 1.8 megawatt solar panel array through a purchase agreement. The solar panels and their maintenance were at no cost to CCMUA and the utility buys power from the contractor at a discounted rate. The solar panels power 10% of energy needs at the wastewater treatment plant and are projected to save $300,000 in energy costs in the first year and $7 million over the life of the 15-year power purchase agreement. The remaining phases of the CCMUA green energy initiative include installing a digester facility by 2016 that would produce enough biogas to meet about 50% to 60% of the utility's power needs. CCMUA also received a $1 million grant from the New Jersey Board of Public Utilities to implement an innovative sewage-to-heat facility which converts latent heat in sewage into heat at the plant. It is expected that this process will be adopted by other large facilities (e.g., hotels) and will have widespread implications across the water-energy industry.

In 2011, CCMUA founded a partnership group called the Camden SMART initiative that integrates water conservation and green infrastructure in order to reduce infiltration/inflow as well as the number of Combined Sewer Overflows (CSOs). Camden SMART is a collaboration of CCMUA, Rutgers University, the New Jersey Department of Environmental Protection and two local non-profit organizations: Cooper's Ferry Partnership and the New Jersey Tree Foundation. These organizations work together to design, build and maintain rain gardens throughout Camden, NJ, including a few on remediated continued on next page

Example of Sustainability Planning *(continued)*

brownfield sites. These rain gardens cost an average of $4.59 per square foot to install and $0.05 per gallon of stormwater diverted from the collection system.

From 2011 to 2014, the Camden SMART initiative planted over 300 new trees and installed 30 rain gardens, capturing approximately 3 million gallons of stormwater per year. CCMUA also received a $5.5 million, low interest loan from the New Jersey Environmental Infrastructure Trust to "daylight" a stream that was previously paved over, convert an abandoned industry building into a riverfront park and construct 10 additional rain gardens. These projects are expected to capture over 30 million gallons of stormwater per year, further reducing the impacts of CSOs. In addition to the work through Camden SMART, CCMUA is reducing CSOs by optimizing the sewer system's performance through changes in operations and maintenance (e.g., cleaning inlets, replacing netting systems, jetlines, etc.) and by pursuing targeted capital replacement and sewer separation.

The Camden SMART initiative and CCMUA have also partnered to establish a water conservation program to alleviate issues associated with reduced water pressure in times of drought and to reduce the burden on the combined sewer system. Camden City has adopted a new water conservation ordinance that limits days and times residents can water lawns and run irrigation systems. To assist customers in conserving water in homes and businesses, CCMUA has distributed water conservation kits and provides conservation tips and information in quarterly bill inserts to increase awareness of conservation practices. At the utility itself, CCMUA is also undertaking an aggressive water loss investigation to reduce potable water loss. Together, these activities reduce water use, which increases the efficiency of CCMUA's operations and reduces costs for customers.

Because climate science is evolving and uncertainty surrounds the timing, nature, direction and magnitude of related impacts, it is important for utilities to continuously assess and respond to new risks and opportunities during the adaptation planning process. This iterative approach has been described for water utilities as part of the Plan-Do-Check-Act approach. This approach is a project-management cycle that can be helpful in promoting continuous improvement by emphasizing evaluation of progress and corrective action when necessary (for more information, see EPA's *Energy Management Guidebook for Wastewater and Water Utilities*).

The National Drinking Water Advisory Council's (NDWAC) Climate Ready Water Utilities (CRWU) Workgroup proposed an adaptive management approach based on a concept similar to Plan-Do-Check-Act. The NDWAC report (2010) outlined steps a utility can take to become more "climate ready." Climate ready water utilities are those drinking water, wastewater and stormwater utilities that are engaged in the process of conducting activities to better understand their climate risks, planning to address climate impacts and implementing adaptation measures to reduce the consequences of climate change. Based on their findings, EPA developed an Adaptive Response Framework to guide utilities through the process of becoming climate ready. Utilities are encouraged to explore each element as part of an adaptive management approach. The Adaptive Response Framework, provides six elements to help utilities build climate readiness: awareness, adaptation, mitigation, policies, community and partnership (**Figure 2.2**).

Figure 2.2. The six elements of the Adaptive Response Framework.

Suggested Actions from the Adaptive Response Framework:

Climate Impacts & Uncertainties:

- Maintain a basic awareness of climate science developments and implications for local operational conditions.
- Encourage utility personnel to examine operating conditions in light of the potential for climate change challenges.
- Conduct screening-level climate impact assessments to identify obvious threats and opportunities.
- Integrate climate impact considerations into normal planning and decision making, including emergency response, capacity and capital planning.

Utility Adaptation & Mitigation Opportunities:

- Understand organizational, operational and capital investment options undertaken by similar utilities to better understand opportunities for no- and low-cost and no-regrets, operational actions and capital investments.
- Expand efforts to identify, understand and evaluate utility climate adaptation and mitigation practices (e.g., enhanced long-range planning methods, hedging strategies and supply and treatment diversification options).

What types of adaptation strategies are included in this Guide?

An effective adaptation plan should include a diverse set of actions that are integrated into other planning efforts, operating practices and infrastructure improvements. Through integrated strategies, utilities can ensure that adaptation actions will address a broad range of challenges while remaining flexible enough to adapt to changing climate conditions and new information. For example, the need for infrastructure improvements can be informed by monitoring conditions, while current emergency response plans can provide resilience until improvements are in place. In this Guide, comparison and prioritization of adaptation options are not provided, although the options are broadly categorized based on the level of effort and relative costs required in their implementation. The three categories of adaptation options included are:

- **Planning strategies**, which include use of models, research, training, supply and demand planning, natural resource management, land use planning and collaboration at watershed and community scales;

- **Operational strategies**, which include efficiency improvements, monitoring, inspections, conservation, demand management and flexible operations; and

- **Capital / infrastructure strategies**, which include construction, water resource diversification, repairs and retrofits, upgrades, phased construction and new technology adoption.

These adaptation options are categorized in terms of the relative anticipated cost of implementation. Planning strategies tend to be relatively less costly than operational and capital strategies; however, there is some diversity in costs within each category. Another factor to consider in assessing costs of options is the potential to recover funds directly (e.g., sell generated power) or avoid costs from inaction (e.g., emergency spending on flood recovery). These benefits effectively reduce the costs of options and need to be explored by each utility as part of the selection and implementation of adaptation options. Three relative cost levels are used in the Guide:

$ Many utilities will try to cope with change by assessing their options to expand operational flexibility to meet the changed operating parameters driven by the climate threat. Costs associated with adaptation options may be minimal.

$$ Some systems can operate beyond design or current capacity without making large changes to the system. Operations and maintenance costs may increase, but would remain less costly than making infrastructure changes.

$$$ After the existing system has reached the limit of its capacity to absorb climate impacts, it becomes necessary to augment or optimize capacity through adoption of new practices and resources. This typically involves a higher level of capital investment.

How do utilities assess adaptation strategies?

Many options exist to address climate change concerns at utilities. When evaluating a response to climate change and assessing potential adaptation strategies, there are several significant issues to be considered such as: deciding which climate information to use, deciding how to incorporate uncertainty and obtaining a better understanding of system capabilities. Several common approaches used by utilities to assess risk and deal with uncertainty in decision-making are described below. In addition, tools have been developed to assess adaptation options in terms of cost and resilience gained (e.g., *Climate Resilience Evaluation and Awareness Tool* or CREAT) as utilities pursue the integration of adaptation into overall capital investment and infrastructure planning.

Assessment approaches

There are many options available to assess how climate change will impact a utility. Three examples of these methods are:

- **Scenario-based** or top-down approaches, which use climate change projection data to inform decision-making. Projections from General Circulation Models (GCMs) are often downscaled to sub-regional spatial scales (tens of kilometers) for impact assessments. This climate information may be coupled with other models (e.g., hydrologic or flood) to predict a system response. Risk is based primarily on the consequences from a damaged or failing system under the scenario being considered (Freas et al. 2008, Brown 2010).

- **Decision-scaling** or threshold-based approaches, which consider how changes in climate will impact performance based on the current capacity of systems. These evaluations will produce thresholds for failure or damage. Risk is gauged based on the likelihood of exceeding thresholds using GCM projections and consequences from a failing or damaged system (Brown 2010).

- **Robust decision-making** approaches, which apply multiple scenarios derived from GCM projections to create ensembles of plausible futures. The performance of adaptation options is considered across these scenarios to identify those options that reduce risk across all or most scenarios and avoid unacceptable outcomes or worst-case scenarios (Lembert and Groves 2010).

Examples of Assessment Approaches:

Inland Empire Utilities Agency

Southern California's Inland Empire Utilities Agency used robust decision-making to evaluate the impacts of climate change on long-term urban water management. The goal was to reject any water strategy that cost more than $3.75 billion. Scenario discovery using 21 climate models and a water management model concluded that the costs would exceed that figure if three things happened concurrently: large precipitation declines, large changes in the price of water imports and reductions in the natural percolation into ground water aquifers. Based on this, a management plan was devised that included: water-use efficiency, capturing storm water for ground water replenishment, water recycling and importing water in wet years so ground water can be extracted in dry years. The Agency found that if all these actions were undertaken, the costs would almost never exceed the $3.75 billion limit (Lembert and Groves 2010).

Sydney Water

Sydney Water (Australia) provides sustainable water, wastewater, recycled water and some stormwater services to more than four million people in Sydney and surrounding areas. Sydney Water addresses climatic and weather-related extreme events through an adaptive management approach, embedded in corporate planning and risk management protocols

Sydney Water has been an early adopter of climate change risk management for its $39 billion worth (Australian dollars) of infrastructure. Over the last 10 years, the organization has considered the impacts of future climate on water supply and demand planning. In collaboration with other state agencies, Sydney Water is addressing the risk of climate impacts by increasing supply diversity including dams, water recycling, water efficiency and desalination.

In 2013, the organization assessed the impacts of future climate on infrastructure, operations and customers through a three year Climate Change Adaptation Program. The three objectives of continued on next page

Examples of Assessment Approaches: *(continued)*

the program were to (1) understand the business impacts related to climate change, including supply chain interdependencies with electricity and telecommunications providers; (2) identify current resilience capabilities to respond to events and (3) cost and prioritize adaptation options for the utility to consider. One of the key outputs from this program is a climate change adaptation quantification tool, called AdaptWater, which calculates both the consequences of climate change hazards and the effectiveness of adaptation options in reducing risk.

Addressing uncertainty and varying capacity to respond at utilities

Because of the great deal of uncertainty surrounding the timing, nature, direction and magnitude of localized climate impacts, it can be a challenge for utilities to address climate change. In particular, it may be difficult to balance climate-related action with current obligations, which requires maintaining service affordability while developing the financial, managerial and technical capacity to meet future needs.

However, this uncertainty should not prevent utilities from taking action now with regards to potential climate change impacts. For some utilities, it is not an option to wait and see or take no action. In fact, the cost of inaction may be greatly underestimated and can be offset by taking preventative action today. Building climate considerations into everyday utility decision making is a current necessity because utility investments are often capital intensive, long-lived and can require long lead times to ensure system reliability and maintenance of desired service levels. Flexible or adaptive management strategies provide a structure for implementing adaptation options for future operations despite these uncertainties and differing capacities.

How can utility laboratories adapt to climate change impacts?

Many projected climate impacts can alter drinking water quality and quantity, which could have implications for environmental and public health laboratories in addition to the utilities themselves. Climate model projections indicate that the frequency and severity of extreme events will increase in many regions of the U.S., including intense precipitation events, prolonged droughts and wildfires. More frequent and more intense flooding events can result in increased sedimentation, turbidity and pollution inputs. Increased surface water temperatures can lead to an increased frequency of algal blooms. Saltwater intrusion into aquifers due to sea-level rise could potentially introduce non-indigenous biological contaminants into water sources. These impacts may challenge laboratories with increased demand for sampling and potential analytical matrix interference issues.

For example, following heavy flooding in Colorado in September 2013, the Colorado Department of Public Health and Environment (CDPHE) was faced with an increased number of samples that required analysis with limited testing supplies to provide to field teams. CDPHE was able to prepare and distribute 40 smaller sampling kits from available supplies to collect flood water samples in the affected area. Volunteers were used to analyze the increased number of samples, and 95% of the data was reported within three working days of the request from the Governor of Colorado.

Extreme flooding events similar to the event in Colorado are projected to occur more frequently due to climate change impacts. Therefore it is important for laboratories supporting water utilities to understand projected climate change impacts and how to adapt to them, increasing their climate continued on next page

Examples of Assessment Approaches: *(continued)*

readiness in the future. The table below contains example adaptation options for environmental and public health laboratories. For more information on how climate change can impact laboratories, review the Water Quality Degradation Group Briefs.

✓	ADAPTATION OPTIONS FOR WATER LABORATORIES	COST
	Develop models to understand potential water quality changes (e.g., increased turbidity and matrix interference) and costs of resultant changes in treatment.	$
	Conduct climate change impacts and adaptation training for personnel.	$
	Participate in *Water Laboratory Alliance (WLA)*-led exercises (i.e., full-scale exercises, tabletop exercises, live tabletop exercise webcasts) and WLA Security Summits to gain insight into emergency response best practices and lessons learned.	—
	Develop emergency response plans and utilize the WLA *Continuity of Operations (COOP) Plan Template* to create standard operating procedures in advance of potential natural disasters.	$
	Improve sample throughput to address surge capacity requirements from increased sampling and analysis needs associated with natural disasters.	$$
	Facilitate development of emergency response field kits with detailed sampling and shipping instructions that can be strategically placed in advance, or quickly dispatched, throughout a state or distribution system.	$$
	Establish alternative power supplies, potentially through on-site generation, to support operations in case of loss of power.	$-$$

Taking Action

When considering which climate-related actions to take now, it is important that utilities develop an understanding of all of the potential benefits of implementing adaptation options beyond increased overall resilience (Danilenko et al. 2010, UKCIP 2011). For example, many options may provide benefits under both current climate conditions and potential future climate conditions. These options are often described as No Regrets options. As used in this Guide, No Regrets describes those adaptation options that provide benefits regardless of future climate conditions. These options would increase resilience to the potential impacts of climate change while yielding other, more immediate economic, environmental or social benefits (WUCA 2010, FAO 2011). However, No Regrets does not mean cost-free; these options still have real or opportunity costs or represent trade-offs that should be considered by utility owners and operators (Wilby 2008, Heltberg et al. 2009). Within the briefs, No Regrets adaptation options are identified with this icon.

Only implementing No Regrets options at a utility may not be enough to build resilience against climate impacts. Other types of actions include those that provide benefits particularly if climate projections become reality (low-regrets or climate-justified) as well as actions that reduce greenhouse gas emissions and provide co-benefits (i.e., energy efficiency, optimization and reduced operating costs).

Example of Utilizing No Regrets Options:

The Tualatin Valley Water District (TVWD) provides an average of 23 million gallons of water per day to more than 200,000 customers in Beaverton, Hillsboro and Tigard, Oregon. Many other stressors, in addition to climate change, present potential risks and uncertainties as related to future water supply and water quality issues. Recognizing this and the uncertainty in climate change projections, TVWD has developed plans that employ No Regrets strategies focused on building a more resilient regional water supply. Actions being pursued include data collection, diversifying supply sources, investigating water system interties and engaging customers in extensive conservation efforts.

As part of its overall strategy, TVWD collaborates with other utilities in the region through the area's Regional Water Providers Consortium (RWPC). The RWPC Strategic Plan identifies the need to encourage partnerships between providers and facilitate and support reliable back-up water supplies for all water providers should any source or transmission facilities become unavailable due to an emergency or natural disaster. Efforts thus far have produced an ArcGIS geodatabase of all existing water system facilities within the region, including existing water system interconnections and a pipe network overlay. As a result of this collaboration, TVWD has identified strategic regional interties and plans to diversify sources by identifying alternate surface water supplies and further evaluating aquifer storage and recovery.

TVWD first established an effective conservation program in 1993. The program has been very successful, and District customers recently reduced water usage from 2005 to 2011 by 13% (in gallons per capita per day), more than doubling its 0.8% per year goal. Admittedly, other factors likely contributed to the reduction, but conservation goals were met primarily through a combination of rebates, free water-efficient hardware, consultations, technical assistance to large water users and outreach to customers. TVWD works to mitigate any negative operational, environmental and societal effects through its Sustainability Program, which provides leadership, education, analysis, project management and accountability for the District's sustainability efforts. Key objectives in pursuing sustainability include reducing TVWD's carbon footprint without compromising customer service, enhancing understanding of customer water usage and demand, generating on-site solar energy and maintaining stewardship of assets. TVWD is using a proactive adaptive management approach to continue meeting its sustainability and resiliency goals. TVWD collects and analyzes utility and regional data on a regular basis in order to ensure the District can meet future time demands and reliability concerns. As new information becomes available, water planning decisions are reassessed and modified.

Where can utilities find more information on adaptation planning and other Climate Ready Water Utilities activities?

Adaptation options found in this Guide provide the building blocks for utility adaptation strategies. Further consideration of these options is required, as described above in the adaptation planning section. See the links below for supporting information on available products and resources available to support adaptation planning.

Supporting Information on EPA Initiatives and Products

| EPA CRWU | CREAT | EPA Green Infrastructure |
| CRWU ARF | EPA Sustainability | WaterSense |

Example Technical and Informational Resources to Support Assessments

| NOAA Flood Watch Data | US Drought Portal | US Global Change Research Program | Bias Corrected & Downscaled WCRP CMIP3 & CMIP5 Climate Projections | EPA BASINS Climate Assessment Tool |

How does a utility get started?

This Guide includes a collection of briefs (see below) that provide summaries of climate change projections, descriptions of specific climate impacts that water utilities may experience and descriptions of suggested adaptation options to address these impacts – with their relative costs. The briefs provide the user with comprehensive information on climate change impacts and adaptation planning. Alternatively, each brief can also be considered a stand-alone resource.

Climate Region Briefs—National and regional descriptions of climate change projections are provided in the Climate Region Briefs. The material in these briefs is drawn from the most recent U.S. Global Change Research Program assessment (2014). These briefs provide an overview of climate change projections in each region, along with associated impacts drinking water and wastewater utilities will face. Clicking on a region will bring you to that particular Climate Region Brief.

LINKS TO CLIMATE REGION BRIEFS

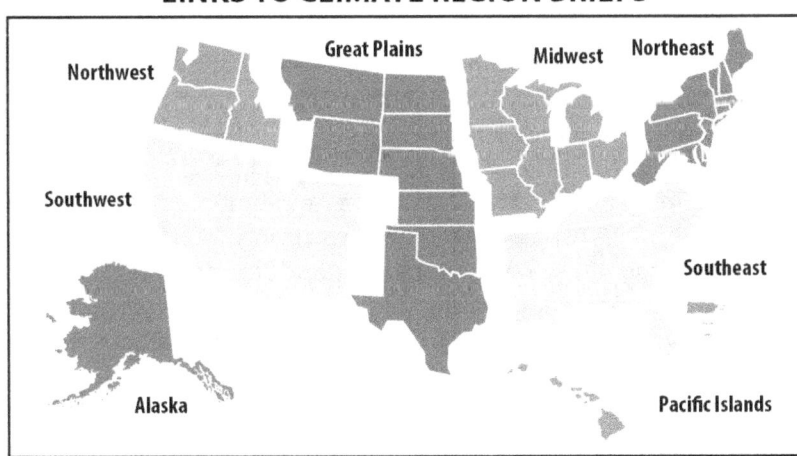

Group and Strategy Briefs —Summaries of general impacts that drinking water and wastewater utilities may face are contained in the Group Briefs, which can be accessed by clicking on an impact group in the table below. These briefs contain a comprehensive list of adaptation options to address a group of similar potential impacts.

LINKS TO STRATEGY BRIEFS

BRIEFS BY GROUP			DW	WW	Sustainability Briefs EM	GI	WDM
Drought		Reduced groundwater recharge	💧				
		Lower lake & reservoir levels	💧			⬤	⬤
		Changes in seasonal runoff & loss of snowpack	💧				
Water Quality Degradation		Low flow conditions & altered water quality		💧			
		Saltwater intrusion into aquifers	💧		⬤	⬤	
		Altered surface water quality	💧	💧			
Floods		High flow events & flooding	💧	💧		⬤	
		Flooding from coastal storm surges	💧	💧			
Ecosystem Changes		Loss of coastal landforms / wetlands	💧	💧		⬤	
		Increased fire risk & altered vegetation	💧	💧			
Service Demand & Use		Volume & temperature challenges	💧	💧			
		Changes in agricultural water demand	💧		⬤	⬤	⬤
		Changes in energy sector needs & energy needs of utilities	💧	💧			

Click on a group name or icon above to read more about these climate impacts or click on a water drop above to read more about a specific impact. Click on a Sustainable Strategy icon to read more about energy management, green infrastructure or water demand management strategies.

The Group Briefs also include links to the more specific Strategy Briefs that provide more detailed information on potential climate change-related impacts for drinking water, wastewater and stormwater utilities. Each Strategy Brief provides general climate information related to the projected impact, options for adaptation strategies to address them, relative cost information and examples describing how a specific utility has implemented at least one of the options listed. Clicking on a water drop (💧) in the table above will bring you to that Strategy Brief. Most briefs apply to either drinking water (DW) or wastewater and stormwater (WW) utilities. In the case of the ecosystem-related impacts and energy sector needs, briefs apply to DW and WW together. Clicking on the energy management (EM), green infrastructure (GI) or water demand management (WDM) icons in the Sustainability Briefs columns will bring you to those respective briefs. **From anywhere in the Guide, you can also return to a previously-viewed page by pressing the ALT key with the left arrow key.**

Austin, J.A., & Colman, S.M. Lake Superior summer water temperatures are increasing more rapidly than regional air temperatures: A positive ice-albedo feedback. *Geophysical Research Letters.* **2007.** 34, L06604, doi: 10.1029/2006GL029021.

Backlund, P., Janetos, A., Schimel, D.S., Hatfield, M.G., Archer, S.R., & Lettenmaier, D. The Effects of Climate Change on Agriculture, Land Resources, Water Resources, and Biodiversity in the United States (SAP 4.3). A Report by the U.S. Climate Change Science Program and the Subcommittee on Global Change Research, U.S. Department of Agriculture, Washington, DC. **2008.** 362 pp.

Bailey, R.G. Description of the Ecoregions of the United States (2nd ed.). U.S. Department of Agriculture Forest Service, Washington, DC. **1995.** 108 pp.

Barnett, T., Malone, R., Pennell, W., Stammer, D., Semtner, B., & Washington, W. The effects of climate change on water resources. *Climatic Change.* **2004.** 62, 1–11.

Boerema, A. Managing the demand for water in Sydney, Waterwise Annual Water Efficiency Conference 2008 – The Road to Water Efficiency in the UK, 8–9 April 2008, Keble College, Oxford University, Oxford, UK. **2008.**

Brown, C. Decision-scaling for robust planning and policy under climate uncertainty. Background paper for World Development Report. The World Bank, Washington, DC. **2011.**

Brown, P.M., Heyerdahl, E.K., Kitchen, S.G., & Weber, M.H. Climate effects on historical fires (1630–1900) in Utah. *International Journal of Wildland Fire.* **2008.** 17, 28-39.

Caldwell, P.V., Sun, G., McNulty, S.G., Cohen, E.C., & Moore Myers, J.A. Impacts of impervious cover, water withdrawals, and climate change on river flows in the Conterminous US. *Hydrology and Earth System Sciences Discussions.* **2012.** 9, 4263-4304.

Cayan, D.R., Das, T., Pierce, D.W., Barnett, T.P., Tyree, M., & Gershunov, A. Future dryness in the southwest US and the hydrology of the early 21st century drought. Proceedings of the National Academy of Sciences. **2010.** 107, 21271-21276.

Cayan, D., Kunkel, K., Castro, C., Gershunov, A., Barsugli, J., Ray, A., . . . Duffy, P. Ch. 6: Future climate: Projected average. In *Assessment of Climate Change in the Southwest United States: A Report Prepared for the National Climate Assessment,* G. Garfin, A. Jardine, R. Merideth, M. Black, & S. LeRoy (Eds.). Island Press. **2013.** 153-196.

Clean Air Partnership (CAP). Cities Preparing for Climate Change: A Study of Six Urban Regions. **2007.**

Climate Adaptation Knowledge Exchange (CAKE). Proactive incorporation of sea-level rise: The case of Deer Island Wastewater Treatment Plant. **2011.** http://www.cakex.org/case-studies/2791 (accessed September 26, 2014).

Cohen, R. The water-energy nexus. *Southwest Hydrologist.* **2007.** 6(5), 16-19.

Colorado Water Conservation Board (CWCB). Joint Front Range Climate Change Vulnerability Study. **2011.** http://cwcb.state.co.us/environment/climate-change/Pages/JointFrontRangeClimateChangeVulnerabilityStudy.aspx (accessed September 26, 2014).

Conrads, A., Covich, A.P., Cruise, J., Feldt, J., Georgakakos, A.P., McNider, R.T., . . . Terando, A. Impacts of climate change and variability on water resources in the Southeast USA. In *Climate of the Southeast United States: Variability, Change, Impacts, and Vulnerability,* K.T. Ingram, K. Dow, L. Carter, & J. Anderson (Eds.). Island Press. **2013.** 210-236.

Danilenko, A., Dickson, E., & Jacobsen, M. Climate Change and Urban Water Utilities: Challenges and Opportunities. Water Working Notes No. 24. The World Bank, Washington, DC. **2010.**

Denver Water. Supply Planning. **2014.** http://www.denverwater.org/SupplyPlanning/WaterSupply/PartnershipUSFS/ (accessed September 26, 2014)

Easterling, W.E., Hurd, B.H., & Smith, J.B. Coping with Global Climate Change: The Role of Adaptation in the United States. The Pew Center on Global Climate Change. **2004.** http://www.c2es.org/docUploads/Adaptation.pdf (accessed September 26, 2014).

Famiglietti, J., Lo, M., Ho, S.L., Bethune, J., Anderson, K.J., Syed, T.H., . . . Rodell, M. Satellites measure recent rates of groundwater depletion in California's Central Valley. *Geophysical Research Letters.* **2011.** 38, L03403.

Fay, M., & Ebinger, J. A Framework for developing adaptation plans. In *Adapting to Climate Change in Eastern Europe and Central Asia,* Fay, M., Block, R., & Ebinger, J. (Eds). The World Bank, Washington, DC. **2010.**

Flannigan, M.D., Stocks, B.J., & Wotton, B.M. Forest fires and climate change. *Science of the Total Environment.* **2000.** 262, 221-230.

Food and Agriculture Organization of the United Nations (FAO). Climate change, water and food security. **2011.** http://www.fao.org/docrep/014/i2096e/i2096e.pdf (accessed September 26, 2014).

Freas, K., Bailey, B., Munevar, A., Butler, S. Incorporating climate change in water planning. *Journal of the American Water Works Association.* **2008.** 100(6), 92–99.

Fritze, H., Stewart, I.T., & Pebesma, E.J. Shifts in Western North American snowmelt runoff regimes for the recent warm decades. *Journal of Hydrometeorology.* **2011.** 12, 989-1006.

Foti, R., Ramirez, J.A., & Brown, T.C. Vulnerability of U.S. Water Supply to Shortage: A Technical Document Supporting the Forest Service 2010 RPA Assessment. RMRS-GTR-295. U.S. Department of Agriculture, Forest Service, Rocky Mountain Research Station, Fort Collins, Colorado. **2012.** 147 pp.

Global Climate Change Impacts in the United States. Karl, T.R., Melillo, J.T., and Peterson, T.C., Eds. Cambridge University Press, **2009.**

Groisman, P.Y., Knight, R.W., Easterling, D.R., Karl, T.R., Hegerl, G.C., & Razuvaev, V.N. Trends in intense precipitation in the climate record. *Journal of Climate.* **2005.** 18(9), 1326–1350.

Groisman, P.Y., Knight, R.W., & Zolina, O.G. Recent trends in regional and global intense precipitation patterns. *Climate Vulnerability,* R.A. Pielke, Sr. (Ed.). Academic Press. **2013.** 25-55.

Groves, D.G., Yates, D., & Tebaldi, C. Developing and applying uncertain global climate change projections for regional water management planning. *Water Resources Research.* **2008.** 44.

Hammar-Klose, E., & Thieler, E. National Assessment of Coastal Vulnerability to Future Sea-Level Rise: Preliminary Results for the US Atlantic, Pacific and Gulf of Mexico Coasts. U.S. Geological Survey. **2001.** US Reports 99–593, 00-178, and 00-179.

Hayhoe, K., et al. Emission pathways, climate change, and impacts on California. *Proceedings of the National Academy of Sciences,* **2004.** 101(34), 12422–12427.

Hayhoe, K., Wake, C., Anderson, B., Liang, X.-Z., Maurer, E., Zhu, J., . . . Wuebbles, D. Regional climate change projections for the Northeast USA. *Mitigation and Adaptation Strategies for Global Change.* **2008.** 13(5-6), 425-436.

Hayhoe, K., Sheridan, S., Greene, J.S., & Kalkstein, L. Climate change, heat waves, and mortality projections for Chicago. *Journal of Great Lakes Research.* **2010.** 36, 65–73.

Heltberg, R., Siegel, P.B., & Jorgensen, S.L. Addressing human vulnerability to climate change: Toward a 'no-regrets' Approach. *Global Environmental Change.* **2009.** 19, 89–99.

Hoerling, M.P., Eischeid, J.K., Quan, X.-W., Diaz, H.F., Webb, R.S., Dole, R.M., & Easterling, D.R. Is a transition to semi-permanent drought conditions imminent in the Great Plains? *Journal of Climate.* **2008.** 25, 8380–8386.

Horton, R., Gornitz, V., Bowman, M., & Blake, R. Climate observations and projections. *Annals of the New York Academy of Sciences.* **2010.** 1196, 41-62.

Intergovernmental Panel on Climate Change (IPCC). Special Report on Emissions Scenarios. A special report of Working Group III of the Intergovernmental Panel on Climate Change. N. Nakićenović, & R. Swart (Eds.). Cambridge University Press. **2000.** http://www.ipcc.ch/ipccreports/sres/emission/index.htm (accessed September 26, 2014).

Intergovernmental Panel on Climate Change (IPCC). Climate Change 2013: The Physical Science Basis. Contribution of Working Group I to the Fifth Assessment Report of the Intergovern¬mental Panel on Climate Change. T.F. Stocker, D. Qin, G.-K. Plattner, M. Tignor, S.K. Allen, J. Boschung, A. Nauels, Y. Xia, V. Bex, & P.M. Midgley (Eds.). Cambridge University Press, Cambridge, United Kingdom and New York, NY. **2013.** 1535 pp.

Johnson, T. Battling saltwater intrusion in the Central and West Basins. *Water Replenishment District of Southern California Technical Bulletin.* **2007.** 13(Fall). http://www.wrd.org/engineering/seawater-intrusion-los-angeles.php (accessed September 26, 2014).

Karl, T.R., Melillo, J.T., & Peterson, T.C. (Eds.). Global Climate Change Impacts in the United States. Cambridge University Press. **2009.** 189 pp.

Karl, T.R., Gleason, B.E., Menne, M.J., McMahon, J.R., Heim, Jr., R.R., Brewer, M.J., . . . D. R. Easterling. U.S. temperature and drought: Recent anomalies and trends. *Eos, Transactions, American Geophysical Union.* **2012.** 93, 473-474.

Keener, V.W., Hamilton, K., Izuka, S.K., Kunkel, K.E., Stevens, L.E., & Sun, L. Regional Climate Trends and Scenarios for the U.S. National Climate Assessment. Part 8. Climate of the Pacific Islands. *NOAA Technical Report NESDIS.* **2012.** 142-8, 44.

Knutson, T.R., McBride, J.L., Chan, J., Emanuel, K., Holland, G., Landsea, C., . . . Sugi, M. Tropical cyclones and climate change. *Nature Geoscience.* **2010.** 3, 157-163.

Kunkel, K.E., Easterling, D.R., Kristovich, D.A.R., Gleason, B., Stoecker, L., & Smith, R. Meteorological causes of the secular variations in observed extreme precipitation events for the conterminous United States. *Journal of Hydrometeorology.* **2012.** 13, 1131–1141.

Kunkel, K.E., Stevens, L.E., Stevens, S.E., Sun, L., Janssen, E., Wuebbles, D., . . . J. G. Dobson.: Regional Climate Trends and Scenarios for the U.S. National Climate Assessment: Part 2. Climate of the Southeast U.S. NOAA Technical Report 142-2. National Oceanic and Atmospheric Administration, National Environmental Satellite, Data, and Information Service, Washington, DC. **2013.** 103 pp.

Lembert, R.J., & Groves, D.G. Identifying and evaluating robust adaptive policy responses to climate change for water management agencies in the American west. *Technological Forecasting and Social Change.* **2010.** 77, 960–974.

Litschert, S.E., Brown, T.C., & Theobald, D.M. Historic and future extent of wildfires in the Southern Rockies Ecoregion, USA. *Forest Ecology and Management.* **2012.** 269, 124-133.

Littell, J. S., McKenzie, D., Peterson, D.L, & Westerling, A.L. Climate and wildfire area burned in western US ecoprovinces, 1916-2003. *Ecological Applications.* **2009.** 19, 1003-1021.

Mackey, S. Great Lakes nearshore and coastal systems. In *U.S. National Climate Assessment Midwest Technical Input Report*, J. Winkler, J. Andresen, J. Hatfield, D. Bidwell, & D. Brown (Eds.). Great Lakes Integrated Sciences and Assessments (GLISA), National Laboratory for Agriculture and the Environment, **2012**. 14. http://glisa.msu.edu/docs/NCA/MTIT_Coastal.pdf (accessed September 26, 2014).

Mansur, E., Mendelsohn, R., & Morrison, W. Climate change adaptation: A study of fuel choice and consumption in the US energy sector. *Journal of Environmental Economics and Management*. **2008**. 55, 175-193.

Meehl, G.A., Covey, C., Delworth, T., Latif, M., McAvaney, B., Mitchell, J.F.B., . . . Taylor, K.E. The WCRP CMIP3 multi-model dataset: a new era in climate change research. *Bulletin of the American Meteorological Society*. **2007**. 88(9), 1383-1394.

Meehl, G.A., Stocker, T.F., Collins, W.D., Friedlingstein, P., Gaye, A.T., Gregory, J.M., . . . Zhao, Z.-C. Global Climate Projections. In *Climate Change 2007: The Physical Science Basis. Contribution of Working Group I to the Fourth Assessment Report of the Intergovernmental Panel on Climate Change* S. Solomon, D. Qin, M. Manning, Z. Chen, M. Marquis, K.B. Avery, M. Tignor, & H.L. Miller (Eds.). Cambridge University Press, Cambridge, United Kingdom and New York, NY. **2007**.

Metropolitan Water District of Southern California (Metropolitan). Integrated Resources Plan (IRP) Update. **2011**. http://www.mwdh2o.com/mwdh2o/pages/yourwater/irp/ (accessed September 26, 2014).

Miller, K., & Yates, D. Climate Change and Water Resources: a Primer for Municipal Water Providers. American Water Works Association. **2005**.

Milly P.C.D., Betancourt, J., Falkenmark, M., Hirsch, R.M., Kundzewicz, Z.W., Lettenmaier, D.P., & Stouffer, R.J. Stationarity is dead: whither water management? *Science*. **2008**. 319 (5863), 573–574.

Milly, P.C.D., Dunne, K.A., & Vecchia, A.V. Global pattern of trends in streamflow and water availability in a changing climate. *Nature*. **2005**. 438(7066), 347–350.

Min, S., Zhang, X., Zwiers, F.W., & Hegerl, G.C. Human contribution to more-intense precipitation extremes. *Nature*. **2011**. 470, 378–381.

National Association of Clean Water Agencies (NACWA). Confronting Climate Change: An Early Analysis of Water and Wastewater Adaptation Costs. **2009**. https://www.nacwa.org/index.php?option=com_content&view=article&id=939&catid=8&Itemid=7 (accessed September 26, 2014).

National Oceanic and Atmospheric Administration's (NOAA) National Climatic Data Center (NCDC). State Climate Extremes Committee - Records. **2012**.

National Oceanic and Atmospheric Administration (NOAA), Great Lakes Environmental Research Laboratory. Coasts, water levels, and climate change: A Great Lakes perspective, 2012. *Climatic Change*. **2013**. 120, 697–711.

National Oceanic and Atmospheric Administration (NOAA). Billion Dollar Weather/Climate Disasters. National Oceanic and Atmospheric Administration. http://www.ncdc.noaa.gov/billions (accessed September 26, 2014).

National Research Council (NRC). Climate Stabilization Targets: Emissions, Concentrations, and Impacts over Decades to Millennia. Committee on Stabilization Targets for Atmospheric Greenhouse Gas Concentration, The National Academies Press, Washington, DC. **2011**. 298 pp.

Nearing, M. A., Jetten, V., Baffaut, C., Cerdan, O., Couturier, A., Hernandez, M., . . . van Oost, K. Modeling response of soil erosion and runoff to changes in precipitation and cover. *Catena*. **2005**. 61, 131-154.

New York City Panel on Climate Change. Climate Change Adaptation in New York City: Building a Risk Management Response. Vol. 1196.New York City (NYC) PlaNYC Report on Water Network. **2010**. 328 pp. http://www.nyas.org/Publications/Annals/Detail.aspx?cid=ab9d0f9f-1cb1-4f21-b0c8-7607daa5dfcc (accessed September 26, 2014).

New York City (NYC) PlaNYC. A Stronger, more resilient New York. **2013.** http://www.nyc.gov/html/sirr/html/report/report.shtml (accessed September 26, 2014).

Ntelekos, A.A., Oppenheimer, M., Smith, J.A., & Miller, A.J. Urbanization, climate change and flood policy in the United States. *Climatic Change.* **2010.** 103, 597–616.

Oats, R., & Webb, D. The London Rivers Action Plan. 13th International Rivers Symposium. Perth, Australia. **2009.** http://www.therrc.co.uk/lrap/lplan.pdf (accessed September 26, 2014).

Obeysekera, J., Irizarry, M., Park, J., Barnes, J., & Dessalegne, T. Climate change and its implications for water resources management in south Florida. *Stochastic Environmental Research and Risk Assessment.* **2011.** 25, 495-516.

Orlowsky, B., & Seneviratne, S.I. Global changes in extreme events: Regional and seasonal dimension, **2012.** *Climatic Change.* 10, 669-696.

Pacific Institute. Improving Water Management through Groundwater Banking: Kern County and the Rosedale-Rio Bravo Water Storage District. **2011.** http://www.pacinst.org/wp-content/uploads/sites/21/2013/02/groundwater_banking3.pdf (accessed September 26, 2014).

Parris, A., Bromirski, P., Burkett, V., Cayan, D., Culver, M., Hall, J., . . . Weiss, J. Global Sea Level Rise Scenarios for the United States National Climate Assessment. NOAA Tech Memo OAR CPO-1. National Oceanic and Atmospheric Administration, Silver Spring, MD. **2012.** 37 pp.

Patz, J.A., Vavrus, S.J., Uejio, C.K., & McLellan, S.L. Climate change and waterborne disease risk in the Great Lakes region of the US. *American Journal of Preventive Medicine.* **2008.** 35, 451-458.

Pryor, S. C., Kunkel, K.E., & Schoof, J.T. Ch. 9: Did precipitation regimes change during the twentieth century? In *Understanding Climate Change: Climate Variability, Predictability and Change in the Midwestern United States,* Indiana University Press. **2009.** 100-112.

Rahmstorf, S. A semi-empirical approach to projecting future sea-level rise. *Science.* **2007.** 315 (5810), 368–370.

Reutter, J.M., Ciborowski, J., DePinto, J., Bade, D., Baker, D., Bridgeman, T.B., . . . Pennuto, C.M. Lake Erie Nutrient Loading and Harmful Algal Blooms: Research Findings and Management Implications. Final Report of the Lake Erie Millennium Network Synthesis Team, Ohio Sea Grant College Program, The Ohio State University, Lake Erie Millennium Network, Columbus, OH. **2011.** 17 pp. http://www.ohioseagrant.osu.edu/_documents/publications/TS/TS-060%2020June2011LakeErieNutrientLoadingAndHABSfinal.pdf (accessed September 26, 2014).

Schoennagel, T., Sherriff, R.L., & Veblen, T.T. Fire history and tree recruitment in the Colorado Front Range upper montane zone: Implications for forest restoration. *Ecological Applications.* **2011.** 21, 2210–2222.

Schoof, J.T., Pryor, S.C., & Suprenant, J. Development of daily precipitation projections for the United States based on probabilistic downscaling. *Journal of Geophysical Research.* **2010.** 115, 1-13.

Skaggs, R., Janetos, T.C., Hibbard, K.A., & Rice, J.S. Climate and Energy-Water-Land System Interactions Technical Report to the U.S. Department of Energy in Support of the National Climate Assessment. Pacific Northwest National Laboratory, Richland, Washington. **2012.** 152 pp.

Sonoma County Water Agency (SCWA). Carbon Free Water by 2015. **2014.** http://www.scwa.ca.gov/carbon-free-water/ (accessed September 26, 2014).

South Florida Water Management District. Climate Change and Water Management in South Florida. Interdepartmental Climate Change Group Report. South Florida Water Management District. **2009.** http://www.miamidade.gov/greenprint/planning/library/milestone_one/climate_and_water.pdf (accessed September 26, 2014).

DISCLAIMER

The Climate Ready Water Utilities Adaptation Strategies Guide for Water Utilities was prepared by the U.S. Environmental Protection Agency (EPA) as an informational resource to assist drinking water and wastewater utility owners in understanding and addressing climate change risks. It does not purport to be a comprehensive or exhaustive list of all impacts and potential risks from climate change.

The information contained in this Guide was developed in accordance with best industry practices. It should not be exclusively relied on in conducting risk assessments or developing response plans. This information is also not a substitute for the professional advice of an attorney or environmental or climate change professional. This information is provided without warranty of any kind and EPA hereby disclaims any liability for damages, arising from the use of the Guide, including, without limitation, direct, indirect or consequential damages including personal injury, property loss, loss of revenue, loss of profit, loss of opportunity or other loss.

South Monmouth Regional Sewerage Authority (SMRSA). Sea Girt Pump Station Mobile Enclosure: Improving the Environmental Infrastructure of a Small Coastal Community, Sea Girt Avenue Pump Station Reconstruction. **2012.** http://www.smrsa.org/whats-new/innovative-designs (accessed September 26, 2014).

Sovacool, B.K., & Sovacool, K.E. Identifying future electricity-water tradeoffs in the United States. *Energy Policy.* **2009.** 37, 2763-2773.

Standish-Lee, P., & Lecina, K. Getting ready for climate change implications for the western USA. *Water Science and Technology.* **2008.** 58 (3), 727-733.

Stewart, B.C., Kunkel, K.E., Stevens, L.E., Sun, L., & Walsh, J.E. Regional Climate Trends and Scenarios for the U.S. National Climate Assessment: Part 7. Climate of Alaska. National Oceanic and Atmospheric Administration Technical Report NESDIS 142-7. **2013.** 60 pp.

Sun, G., Arumugam, S., Caldwell, P.V., Conrads, P.A., Covich, A.P., Cruise, J., . . . Terando, A. Impacts of climate change and variability on water resources in the Southeast USA. In *Climate of the Southeast United States: Variability, Change, Impacts, and Vulnerability,* K.T. Ingram, K. Dow, L. Carter, & J. Anderson (Eds.). Island Press. **2013.** 210-236.

Trumpickas, J., Shuter, B.J., & Minns, C.K. Forecasting impacts of climate change on Great Lakes surface water temperatures. *Journal of Great Lakes Research.* **2009.** 35, 454-463.

Tucson Water. 2008 Update to Water Plan: 2000-2050. **2008.** http://www.tucsonaz.gov/water/waterplan-2008 (accessed September 26, 2014).

UK Climate Impacts Program (UKCIP). AdOpt: Identifying adaptation options. **2011.** http://www.ukcip.org.uk/wordpress/wp-content/PDFs/ID_Adapt_options.pdf (accessed September 26, 2014).

UK Department of Environment. Thames Estuary 2100 Plan. **2011.** https://www.gov.uk/government/publications/thames-estuary-2100-te2100/thames-estuary-2100-te2100 (accessed September 26, 2014).

UNEP. Climate Change in the Caribbean and the Challenge of Adaptation. United Nations Environment Programme, Regional Office for Latin America and the Caribbean. **2008.** 92 pp.

University of Alaska Fairbanks. Permafrost Lab. Geophysical Institute.

U.S. Climate Change Science Program. The Effects of Climate Change on Agriculture, Land Resources, Water Resources, and Biodiversity. U.S. Environmental Protection Agency, **2008.** 362 pp.

U.S. Department of Energy (DOE). Energy Demands on Water Resources. Report to Congress on the Interdependency of Energy and Water. **2006.**

U.S. Department of Interior (DOI) Bureau of Reclamation. Colorado River Basin Water Supply and Demand Study. **2012.** http://www.usbr.gov/lc/region/programs/crbstudy/finalreport/index.html (accessed September 26, 2014).

U.S. Energy Information Administration. Annual Energy Outlook 2010 with Projections to 2035. **2010.**

U.S. Environmental Protection Agency (EPA). Climate Change Vulnerability Assessments: Four Case Studies of Water Utility Practices. EPA-600-R-10-077F. **2010a.**

U.S. Environmental Protection Agency (EPA). Green Infrastructure Case Studies: Municipal Policies for Managing Stormwater with Green Infrastructure. EPA-841-F-10-004. **2010b.**

U.S. Environmental Protection Agency (EPA). Climate Change Adaptation for Maryland Water Utilities. **2012.** http://www.epa.gov/reg3wapd/pdf/pdf_drinking/20120516_CCbrochure_Web.pdf (accessed September 26, 2014).

U.S. Geological Survey (USGS). Summary of Estimated Water Use in the United States in 2005. **2009.** http://pubs.usgs.gov/fs/2009/3098/ (accessed September 26, 2014).

U.S. Global Change Research Program (USGCRP). Weather and Climate Extremes in a Changing Climate. Regions of Focus: North America, Hawaii, Caribbean and US Pacific Islands. T.R. Karl, G.A. Meehl, C.D. Miller, S.J. Hassol, A.M. Waple, & W.L. Murray (Eds.). Cambridge University Press. **2008.**

U.S. Global Change Research Program (USGCRP). Global Climate Change Impacts in the United States. T.R. Karl, J.M. Melillo, & T.C. Peterson (Eds.). Cambridge University Press. **2009.**

U.S. Global Change Research Program (USGCRP). Third National Climate Assessment. **2014.** http://nca2014.globalchange.gov/ (accessed September 26, 2014).

Vokral, J., Gumb, D., & Mehrotra, S. Staten Island Bluebelt Program: A Natural Solution to Environmental Problems. *Stormwater*. **2001.** 2. http://www.stormh2o.com/SW/Articles/Staten_Island_Bluebelt_Program_A_Natural_Solution_3321.aspx (accessed October 27, 2014).

Wallis, M.J., Ambrose, M.R., & Chan, C.C. Climate Change: charting a watercourse in an uncertain future. *Journal of the American Water Works Association*. **2008.** 100 (6), 70–79.

Water Utility Climate Alliance (WUCA). Decision Support Planning Methods: Incorporating Climate Change Uncertainties into Water Planning. **2010.** http://www.wucaonline.org/assets/pdf/pubs_whitepaper_012110.pdf (accessed September 26, 2014).

Wang, J., Bai, X., Hu, H., Clites, A., Colton, M., & Lofgren, B. Temporal and spatial variability of Great Lakes ice cover, 1973-2010. *Journal of Climate*. **2012.** 25, 1318-1329.

Westerling, A.L., Gershunov, A., Brown, T.J., Cayan, D.R., & Dettinger, M.D. Climate and wildfire in the western United States. *Bulletin of the American Meteorological Society*. **2003.** 84, 595-604.

Westerling, A.L., Hidalgo, H.G., Cayan, D.R., & Swetnam, T.W. Warming and earlier spring increase western U.S. forest wildfire activity. *Science*. **2006.** 313, 940-943.

Westerling, A., Bryant, B., Preisler, H., Holmes, T., Hidalgo, H., Das, T., & Shrestha, S. Climate change and growth scenarios for California wildfire. *Climatic Change*. **2012.** 109, 1-19.

Western Resource Advocates. Protecting the Lifeline of the West: How Climate and Clean Energy Policies Can Safeguard Water. **2010.**

Wehner, M. Changes in daily precipitation and surface air temperature extremes in the IPCC AR4 models. *US CLIVAR Variations*. **2005.** 3(3), 5–9.

Wehner, M. F. Very extreme seasonal precipitation in the NARCCAP ensemble: Model performance and projections. *Climate Dynamics*. **2013.** 40, 59-80.

Wilby, R.L. Dealing with Uncertainties of Future Climate: The Special Challenge of Semi-Arid Regions. Paper presented at Expo 2008 in Zaragoza, Spain. **2008.** http://www.zaragoza.es/contenidos/medioambiente/cajaAzul/5S1-P3-WilbyACC.pdf (accessed September 26, 2014).

Zimmerman, R., & Faris, C. Infrastructure impacts and adaptation challenges. *Annals of the New York Academy of Sciences*. **2010.** 1196, 63-86.

ADDITIONAL RESOURCES

Interagency Climate Change Adaptation Task Force. National Action Plan: Priorities for Managing Freshwater Resources in a Changing Climate DRAFT. U.S. Executive Office of the President - Council on Environmental Quality. **2011.** http://www.whitehouse.gov/sites/default/files/microsites/ceq/napdraft6_2_11_final.pdf (accessed September 26, 2014).

National Research Council. Warming World Impacts by Degree. Based on: Climate Stabilization Targets: Emissions, Concentrations, and Impacts Over Decades to Millennia. **2011.** http://dels.nas.edu/resources/static-assets/materials-based-on-reports/booklets/warming_world_final.pdf (accessed September 26, 2014).

National Research Council. America's Climate Choices. The National Academies Press, Washington, DC. **2011.** http://www.nap.edu/catalog.php?record_id=12781#toc (accessed September 26, 2014).

Olsen, J.R., Kiang, J., & Waskom, R. (Eds). Workshop on Nonstationarity, Hydrologic Frequency Analysis, and Water Management. Colorado Water Institute Information Series No. 109. **2010.** http://www.cwi.colostate.edu/publications/is/109.pdf (accessed September 26, 2014).

U.S. Environmental Protection Agency. Climate Change Vulnerability Assessments: A Review of Water Utility Practices. EPA 800-R-10-001. **2010.** http://water.epa.gov/scitech/climatechange/upload/Climate-Change-Vulnerability-Assessments-Sept-2010.pdf (accessed September 26, 2014).

U.S. Environmental Protection Agency. Proceedings of the First National Expert and Stakeholder Workshop on Water Infrastructure Sustainability and Adaptation to Climate Change. EPA-600-R-09-010. **2009.** http://www.epa.gov/nrmrl/wswrd/wq/wrap/pdf/workshop/600r09010.pdf (accessed September 26, 2014).

U.S. Environmental Protection Agency. National Water Program Strategy: Response to Climate Change. EPA 850-K-12-004. **2012.** http://water.epa.gov/scitech/climatechange/upload/epa_2012_climate_water_strategy_full_report_final.pdf. (accessed September 26, 2014).

GLOSSARY

Return to Introduction

This glossary provides additional explanation of the adaptation options listed in the strategy briefs provided in this Guide. Each option includes general descriptions of actions that may be taken or clarification of terminology. Pursuit of many of these options may require collaboration with other utilities, local or federal government agencies, other sectors (e.g., energy and agriculture) and the academic community. The options are grouped into categories of similar adaptation strategies, including:

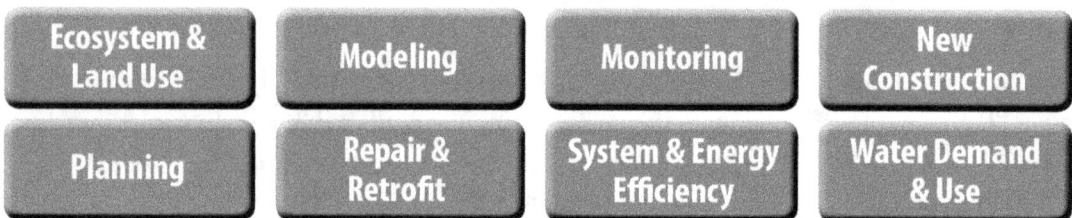

| Ecosystem & Land Use | Modeling | Monitoring | New Construction |
| Planning | Repair & Retrofit | System & Energy Efficiency | Water Demand & Use |

Each option description includes a measure of relative cost, from $ to $$$ (see Page 8 in the Introduction for a description of this scale).

 No Regrets options are marked with this icon. These adaptation options provide benefits regardless of future climate conditions and would increase resilience to the potential impacts of climate change while yielding other, more immediate, economic, environmental or social benefits (WUCA 2010; FAO 2011).

ECOSYSTEM & LAND USE

 Acquire and manage ecosystems ($$$)–Intact natural ecosystems have many benefits for utilities: reducing sediment and nutrient inputs into source water bodies, regulating runoff and streamflow, buffering against flooding and reducing storm surge impacts and inundation on the coasts (e.g., mangroves, saltwater marshes, wetlands). Utilities can also work with regional floodplain managers and appropriate stakeholders to explore non-structural flood management techniques in the watershed. Protecting, acquiring and managing ecosystems in buffer zones along rivers, lakes, reservoirs and coasts can be cost-effective measures for flood control and water quality management.

 Implement green infrastructure on site and in municipalities ($-$$$)–Green infrastructure can help reduce runoff and stormwater flows that may otherwise exceed system capacity. Examples of green infrastructure include: bio-retention areas (rain gardens), low impact development methods, green roofs, swales (depressions to capture water) and the use of vegetation or pervious materials instead of impervious surfaces.

Implement watershed management ($$)–Watershed management includes a range of policy and technical measures. These generally focus on preserving or restoring vegetated land cover in a watershed and managing stormwater runoff. These changes help mimic natural watershed hydrology, increasing groundwater recharge, reducing runoff and improving the quality of runoff.

Integrate flood management and modeling into land use planning ($)–It is critical that future water utility infrastructure be planned and built in consideration of future flood risks. Infrastructure can be built in areas that do not have a high risk of future flooding. Alternately, appropriate flood management plans can be implemented that involve 'soft' adaptation measures such as conserving natural ecosystems or 'hard' measures such as dikes and flood walls.

Study response of nearby wetlands to storm surge events ($)–Coastal wetlands act as buffers to storm surge. Protecting and understanding the ability of existing wetlands to provide protection for coastal infrastructure in the future is important considering projected sea-level rise and possible changes in storm severity.

Update fire models and practice fire management plans ($-$$)–Fire frequency and severity may change in the future, therefore it is important to develop, practice and regularly update management plans to reduce fire risk. Controlled burns, thinning and weed and invasive plant control help to reduce risk in wildfire-prone areas.

MODELING

Conduct extreme precipitation events analyses ($-$$)–An increase in the magnitude or frequency of extreme events can severely challenge water utility systems that were not designed to withstand intense events. Extreme event analyses or modeling can help develop a better understanding of the risks and consequences associated with these types of events.

Conduct sea-level rise and storm surge modeling ($)–Modeling sea-level rise and storm surge dynamics will better inform the placement and protection of critical infrastructure. Generic models have been developed to consider subsidence, global sea-level rise and storm surge effects on inundation, including National Oceanic and Atmospheric Administration's (NOAA) SLOSH (Sea, Lake and Overland Surges from Hurricanes) Model and The Nature Conservancy's Coastal Resilience Tool, amongst others.

 Develop models to understand potential water quality changes ($-$$$)–In many areas, increased water temperatures will cause eutrophication and excess algal growth, which will reduce drinking water quality. The quality of drinking water sources may also be compromised by increased sediment or nutrient inputs due to extreme storm events. These impacts may be addressed with targeted watershed management plans.

 Model and monitor groundwater conditions ($)–Understanding and modeling groundwater conditions will inform aquifer management and projected water quantity and quality changes. Monitoring data for aquifer water level, changes in chemistry and detection of saltwater intrusion can be incorporated into models to predict future supply. Climate change may lead to diminished groundwater recharge in some areas because of reduced precipitation and decreased runoff.

Model and reduce inflow/infiltration in the sewer system ($-$$$)–More extreme storm events will increase the amount of wet weather infiltration and inflow into sanitary and combined sewers. Sewer models can estimate the impact of those increased wet weather flows on wastewater collection system and treatment plant capacity and operations. Potential system modifications to reduce those impacts include infiltration reduction measures, additional collection system capacity, offline storage or additional peak wet weather treatment capacity.

Use hydrologic models to project runoff and future water supply ($)–In order to understand how climate change may impact future water supply and water quality, hydrologic models, coupled with projections from climate models, must be developed. It is important to work towards an understanding of how both the mean and temporal (seasonal) distribution of surface water flows may change. Groundwater recharge, snowpack and the timing of snowmelt are critical areas that may be severely impacted by climate change and should be incorporated into the analysis.

MONITORING

Conduct stress testing on wastewater treatment biological systems to assess tolerance to heat ($$)–Increased surface water temperature may require changes to wastewater treatment systems, as microbial species used may react differently in warmer environments. Stress testing involves subjecting biological systems or bench-top simulations of systems to elevated temperatures and monitoring the impacts on treatment processes.

Manage reservoir water quality ($$)–Changes in precipitation and runoff timing, coupled with higher temperatures due to climate change, may lead to diminished reservoir water quality. Reservoir water quality can be maintained or improved by a combination of watershed management, to reduce pollutant runoff and promote groundwater recharge and reservoir management methods, such as lake aeration.

 Monitor and inspect the integrity of existing infrastructure ($-$$)–Monitoring is a critical component of establishing a measure of current conditions, detecting deterioration in physical assets and evaluating when the necessary adjustments need to be made to prolong infrastructure lifespan.

 Monitor current weather conditions ($)–A better understanding of weather conditions provides a utility with the ability to recognize possible changes in climate change and then identify the subsequent need to alter current operations to ensure resilient supply and services. Observations of precipitation, temperature and storm events are particularly important for improving models of projected water quality and quantity.

 Monitor flood events and drivers ($)–Understanding and modeling the conditions that result in flooding is an important part of projecting how climate change may drive change in future flood occurrence. Monitoring data for sea level, precipitation, temperature and runoff can be incorporated into flood models to improve predictions. Current flood magnitude and frequency of storm events represents a baseline for considering potential future flood conditions.

 Monitor surface water conditions ($)–Understanding surface water conditions and the factors that alter quantity and quality is an important part of projecting how climate change may impact water resources. Monitoring data for discharge, snowmelt, reservoir or stream level, upstream runoff, streamflow, in-stream temperature and overall water quality can be incorporated into models of projected supply or receiving water quality.

Monitor vegetation changes in watersheds ($)–Changes in vegetation alter the runoff that enters surface water bodies and the risk of wildfire to facilities within the watershed. Monitoring vegetation changes can be conducted by ground cover surveys, aerial photography or by relying on the research from local conservation groups and universities.

NEW CONSTRUCTION

Build flood barriers to protect infrastructure ($$-$$$)–Flood barriers to protect critical infrastructure include levees, dikes and seawalls. A related strategy is flood proofing, which involves elevating critical equipment or placing it within waterproof containers or foundation systems.

Build infrastructure needed for aquifer storage and recovery ($$$)–Increasing the amount of groundwater storage available promotes recharge when surface water flows are in excess of demand, thus increasing climate resilience for seasonal or extended periods of drought, and taking advantage of seasonal variations in surface water runoff. Depending on whether natural or artificial aquifer recharge is employed, the required infrastructure may include percolation basins and injection wells.

 Diversify options for water supply and expand current sources ($$-$$$)–Diversifying sources helps to reduce the risk that water supply will fall below water demand. Examples of diversified source water portfolios include using a varying mix of surface water and groundwater, employing desalination when the need arises and establishing water trading with other utilities in times of water shortages or service disruption.

 Increase water storage capacity ($$-$$$)–Increased drought can reduce the safe yield of reservoirs. To reduce this risk, increases in available storage can be made. Methods for accomplishing this may include raising a dam, practicing aquifer storage and recovery, removing accumulated sediment in reservoirs or lowering water intake elevation.

Install low-head dam for saltwater wedge and freshwater pool separation ($$$)–Rising sea levels, combined with reductions in freshwater runoff due to drought, will cause the salt water-freshwater boundary to move further upstream in tidal estuaries. Upstream shifts of this boundary can reduce the water quality of surface water resources. Installation of low-head dams across tidal estuaries can prevent this upstream movement.

 Plan and establish alternative or on-site power supply ($-$$)–Water utilities are one of the major consumers of electricity in the United States. With future electricity demand forecasted to grow, localized energy shortages may occur. The development of "off-grid" sources can be a good hedging strategy for electricity shortfalls. Moreover, redundant power supply can provide resiliency for situations in which natural disasters cause power outages. On-site sources can include solar, wind, inline microturbines and biogas (i.e., methane from wastewater treatment). New and back-up electrical equipment should be located above potential flood levels.

Relocate facilities to higher elevations ($$$)–Relocating utility infrastructure, such as treatment plants and pump stations, to higher elevations would reduce risks from coastal flooding and exposure as a result of coastal erosion or wetland loss.

PLANNING

Adopt insurance mechanisms and other financial instruments ($)–Adequate insurance can insulate utilities from financial losses due to extreme weather events, helping to maintain financial sustainability of utility operations.

Conduct climate change impacts and adaptation training ($)–An important step in developing an adaptation program is educating staff on climate change. Staff should have a basic understanding of the projected range of changes in temperature and precipitation, the increase in the frequency and magnitude of extreme weather events for their region and how these changes may affect the utility's assets and operations. Preparedness from this training can improve utility management under current climate conditions as well.

Develop coastal restoration plans ($-$$)–Coastal restoration plans may protect water utility infrastructure from damaging storm surge by increasing protective habitat of coastal ecosystems such as mangroves and wetlands. Restoration plans should consider the impacts of sea-level rise and development on future ecosystem distribution. Successful strategies may also consider rolling easements and other measures identified by EPA's Climate Ready Estuaries program.

 Develop emergency response plans ($)–Emergency response plans (ERPs) outline activities and procedures for utilities to follow in case of an incident, from preparation to recovery. Some of the extreme events considered in ERPs may change in their frequency or magnitude due to changes in climate, which may require making changes to these plans to capture a wider range of possible events.

 Develop energy management plans for key facilities ($)–Energy management plans identify the most critical systems in a facility, provide backup power sources for those systems and evaluate options to reduce power consumption by upgrading to more efficient equipment. Utilities may develop plans to produce energy, reduce use and work toward net-zero goals.

 Establish mutual aid agreements with neighboring utilities ($)–Beyond the establishment of water trading in times of water shortages or service disruptions, these agreements involve the sharing of personnel and resources in times of emergency (e.g., natural disasters).

 Identify and protect vulnerable facilities ($-$$)–Operational measures to isolate and protect the most vulnerable systems or assets at a utility should be considered. For example, critical pump stations would include those serving a large population and those located in a flood zone. Protection of these assets would then be prioritized based on the likelihood of flood damage and the consequence of service disruption.

Integrate climate-related risks into capital improvement plans ($)–Plans to build or expand infrastructure should consider the vulnerability of the proposed locations to inland flooding, sea-level rise, storm surge and other impacts associated with climate change.

 Participate in community planning and regional collaborations ($-$$)–Effective adaptation planning requires the cooperation and involvement of the community. Water utilities will benefit by engaging in climate change planning efforts with local and regional governments, electric utilities and other local organizations.

Update drought contingency plans ($)–Drought leads to severe pressures on water supply. Drought contingency plans would include the use of alternate water supplies and the adoption of water use restrictions for households, businesses and other water users. These plans should be updated regularly to remain consistent with current operations and assets.

REPAIR & RETROFIT

Implement policies and procedures for post-flood and/or post-fire repairs ($)–Post-disaster policies should minimize service disruption due to damaged infrastructure. These contingency plans should be incorporated into other planning efforts and updated regularly to remain consistent with any changes in utility services or assets.

Implement saltwater intrusion barriers and aquifer recharge ($$$)–As sea level rises, saltwater may intrude into coastal aquifers, resulting in substantially higher treatment costs. The injection of fresh water into aquifers can help to act as a barrier, while intrusion recharges groundwater resources.

Improve pumps for backflow prevention ($$)–Sea-level rise and coastal storm surge can cause wastewater outlets to backflow. To prevent this, stronger pumps may be necessary.

 Increase capacity for wastewater and stormwater collection and treatment ($$$)–Precipitation variability will increase in many areas. Even in areas where precipitation and runoff may decrease on average, the distribution of rainfall patterns (i.e., intensity and duration) can change in ways that impact water infrastructure. In particular, more extreme storms may overwhelm combined wastewater and stormwater systems.

 Increase treatment capabilities ($$$)–Existing water treatment systems may be inadequate to process water of significantly reduced quality. Significant improvement to existing treatment processes or implementation of additional treatment technologies may be necessary to ensure that quality of water supply (or effluent) continues to meet standards as climate change impacts source or receiving water quality.

Install effluent cooling systems ($-$$)–Higher surface temperatures may make meeting water quality standards and temperature criteria more difficult. Therefore, to reduce the temperature of treated wastewater discharges, additional effluent cooling systems may be needed.

Retrofit intakes to accommodate lower flow or water levels ($$-$$$)–In areas where streamflow declines due to climate change, water levels may fall below intakes for water treatment plants.

SYSTEM & ENERGY EFFICIENCY

Finance and facilitate systems to recycle water ($$-$$$)–Recycling greywater frees up more finished water for other uses, expanding supply and decreasing the need to discharge into receiving waters. Receiving water quality limitations may increase due to more frequent droughts. Therefore, to limit wastewater discharges, use of reclaimed water in homes and businesses should be encouraged.

 Improve energy efficiency and optimization of operations ($-$$$)–Water utilities are one of the major consumers of electricity in the United States. With future electricity demand forecasted to grow, localized energy shortages may be experienced. Energy efficiency measures will save in energy costs and make utilities less vulnerable to electricity shortfalls due to high demand or service disruptions from natural disasters.

Practice conjunctive use ($$-$$$)–Conjunctive use involves the coordinated, optimal use of both surface water and groundwater, both intra- and inter-annually. Aquifer storage and recovery is a form of conjunctive use. For example, a utility may store some fraction of surface water flows in aquifers during wet years and withdraw this water during dry years when the river flow is low. Depending on whether natural or artificial aquifer recharge is employed, the required infrastructure may include percolation basins and injection wells.

WATER DEMAND & USE

Encourage and support practices to reduce water use at local power plants ($-$$$)–The electricity sector withdraws the greatest amount of water in the United States, compared with other sectors. Any efforts to reduce water usage by utilities (e.g., closed-loop water circulation systems or dry cooling for the turbines) will increase available water supply. For example, utilities may provide reclaimed water to electric utilities for electricity generation.

Model and reduce agricultural and irrigation water demand ($-$$$)–Agriculture represents the second largest user of water in the United States in terms of withdrawals. In order to forecast and plan for future water supply needs, agricultural (irrigation) demand must be projected, particularly in drought-prone areas. For example, to reduce agricultural water demand, utilities can work with farmers to adopt advanced micro-irrigation technology (e.g., drip irrigation).

Model future regional electricity demand ($)–The electricity sector represents the largest user of water in the United States in terms of withdrawals. In order to forecast future water supply needs, changes in electricity demand related to climate change must be projected.

 Practice water conservation and demand management ($-$$)–An effective and low-cost method of meeting increased water supply needs is to implement water conservation programs that will cut down on waste and inefficiencies. Public outreach is an essential component of any water conservation program. Outreach communications typically include: basic information on household water usage, the best time of day to undertake water-intensive activities and information on and access to water-efficient household appliances such as low-flow toilets, showerheads and front-loading washers. Education and outreach can also be targeted to different sectors (i.e., commercial, institutional, industrial, public sectors). Effective conservation programs in the community include those that provide rebates or help install water meters, water-conserving appliances, toilets and rainwater harvesting tanks.

This page left intentionally blank

This page left intentionally blank

 United States
Environmental Protection
Agency

Adaptation Strategies Guide for Water Utilities

WORKSHEET FOR ADAPTATION PLANNING

This adaptation planning worksheet is provided to help identify and organize adaptation options of interest. Either (1) print this worksheet and fill in the fields by hand while browsing through the Guide or (2) type in the fields electronically and make sure to print or save this worksheet before closing the Guide. There is a completed sample version for your reference following this worksheet.

Contact and Utility Information

Name

Utility Name

Phone

Email

Utility Type   DW ☐   WW ☐   SW ☐

Climate Region   Coasts ☐

Climate-Related Impacts

Review the briefs for your climate region and select those that are of concern to your utility.

Drought
- ☐ Reduced groundwater recharge
- ☐ Lower lake & reservoir levels
- ☐ Changes in seasonal runoff & loss of snowpack

Water Quality Degradation
- ☐ Low flow conditions & altered water quality
- ☐ Saltwater intrusion into aquifers
- ☐ Altered surface water quality

Floods
- ☐ High flow events & flooding
- ☐ Flooding from coastal storm surges

Ecosystem Changes
- ☐ Loss of coastal landforms / wetlands
- ☐ Increased fire risk & altered vegetation

Service Demand & Use
- ☐ Volume & temperature challenges
- ☐ Changes in agricultural water demand
- ☐ Changes in energy sector needs
- ☐ Changes in energy needs of utilities

Sustainability
- ☐ Energy management
- ☐ Green infrastructure
- ☐ Water demand management

List the critical threshold conditions (e.g., specific flood heights, drought durations and peak influent volumes that exceed your current operating capacity) that may result in damage or loss to your assets and water resources. For example, if your previous experience indicates that a daily rainfall total of 3 inches would flood critical pump stations, then document this type of event as a threshold to consider during adaptation planning.

Note specific utility assets and water resources where any damage or loss would impair meeting your utility's mission.

WORSHEET _____

Review the briefs for selected climate impacts and note the adaptation options that you would consider implementing to reduce the consequences of climate change at your utility.

Review the Sustainability Briefs and their individual sustainable practices—list any practices that complement or overlap with options you noted above.

Communication with other utilities—what climate change-related actions have other drinking water and wastewater utilities in your area taken?

Adaptation Implementation Planning _____

Note any potential barriers to implementation, collaborators, performance metrics and any other relevant planning details.

Planning priorities (select)

☐ Timing of action	☐ Sustainability	☐ Available funding
☐ Vulnerability assessment	☐ Energy savings	☐ Other:
☐ Assets impacted	☐ Cost savings	

Use the information documented in this worksheet as a preliminary step in the adaptation planning process. As you continue to monitor conditions and begin implementing adaptation options, revisit the Guide and revise this worksheet accordingly to inform future planning efforts.

Adaptation Strategies Guide for Water Utilities
WORKSHEET FOR ADAPTATION PLANNING SAMPLE

This adaptation planning worksheet is provided to help identify and organize adaptation options of interest. Either (1) print this worksheet and fill in the fields by hand while browsing through the Guide or (2) type in the fields electronically and make sure to print or save this worksheet before closing the Guide.

Contact and Utility Information

Name Dan Frialini

Phone 708-555-1212

Email dfrialini@bcwu.org

Utility Name

Big Creek Water Utility

Utility Type   DW ▣   WW ▣   SW ▣

Climate Region Midwest (IL)   **Coasts** ☐

Climate-Related Impacts

Review the briefs for your climate region and select those that are of concern to your utility.

Drought
- ☐ Reduced groundwater recharge
- ▣ Lower lake & reservoir levels
- ☐ Changes in seasonal runoff & loss of snowpack

Water Quality Degradation
- ☐ Low flow conditions & altered water quality
- ☐ Saltwater intrusion into aquifers
- ▣ Altered surface water quality

Floods
- ▣ High flow events & flooding
- ☐ Flooding from coastal storm surges

Ecosystem Changes
- ☐ Loss of coastal landforms / wetlands
- ▣ Increased fire risk & altered vegetation

Service Demand & Use
- ☐ Volume & temperature challenges
- ☐ Changes in agricultural water demand
- ☐ Changes in energy sector needs
- ▣ Changes in energy needs of utilities

Sustainability
- ▣ Energy management
- ▣ Green infrastructure
- ▣ Water demand management

List the critical threshold conditions (e.g., specific flood heights, drought durations and peak influent volumes that exceed your current operating capacity) that may result in damage or loss to your assets and water resources. For example, if your previous experience indicates that a daily rainfall total of 3 inches would flood critical pump stations, then document this type of event as a threshold to consider during adaptation planning.

* 100-year flood would damage storage tanks
* Creek level drops below current intake would restrict supply
* 50% extent of forest loss from fire would lead to increased erosion from forest into Big Creek
* Water demand reduction target of 15% per capita in 10 years to accommodate population growth
* Targeting reduction in energy use (25%) and net greenhouse gas emissions (50%) in 10 years

Note specific utility assets and water resources where any damage or loss would impair meeting your utility's mission.

Storage tanks: past algal blooms have contaminated tanks and storm-related flood damage
Watershed: fires lead to increases in sediment and nutrient runoff into source waters
Treatment plants: energy costs and recent power outages led to efforts to reduce energy needs and increase on-site generation

Review the briefs for selected climate impacts and note the adaptation options that you would consider implementing to reduce the consequences of climate change at your utility.

Already in place: climate change training for personnel / flood models and temporary flood barrier / weather monitoring and demand reduction and modeling efforts.

To evaluate: new levee, wetlands for flood protection, green infrastructure in the community, collaborative land-use planning project, wildfire surveillance, improved supply-demand models, increased storage and watershed management strategies.

Review the Sustainability Briefs and their individual sustainable practices—list any practices that complement or overlap with options you noted above.

Already in place: energy management plans, water conservation plans and outreach and community rain gardens and downspout disconnect incentive programs.

To evaluate: runoff control buffers in fire-prone areas, additional stormwater retention projects, energy and heat recovery practices, cogeneration project and upgrade to more fuel-efficient vehicle fleet.

Communication with other utilities—what climate change-related actions have other drinking water and wastewater utilities in your area taken?

Other Midwestern utilities have been successful in using wildfire surveillance in cooperation with U.S. Forest Service to limit losses. Representatives planning to attend upcoming utility management conference and joining city-wide flood preparedness task force. Negotiating partnership with local power supplier to facilitate on-site generation and combined public outreach campaign for water and power conservation.

Adaptation Implementation Planning ——————————————————————————————

Note any potential barriers to implementation, collaborators, performance metrics and any other relevant planning details.

Budget available for next decade / limited space for expansion of facilities / potential for relocation of facilities unknown. Watershed for Big Creek, including Big Creek Forest, and paper mill. Watershed managers heard about the ongoing BCWU climate assessment and wanted to know if the utility was seeking input or collaboration opportunities. Others include regional assessment team, City of Cicero, City of Chicago and Big Creek Defenders (local advocacy group).

Planning priorities (select)

■ Timing of action	■ Sustainability	■ Available funding
☐ Vulnerability assessment	■ Energy savings	☐ Other:
■ Assets impacted	☐ Cost savings	

Use the information documented in this worksheet as a preliminary step in the adaptation planning process. As you continue to monitor conditions and begin implementing adaptation options, revisit the Guide and revise this worksheet accordingly to inform future planning efforts.

EPA United States
Environmental Protection
Agency

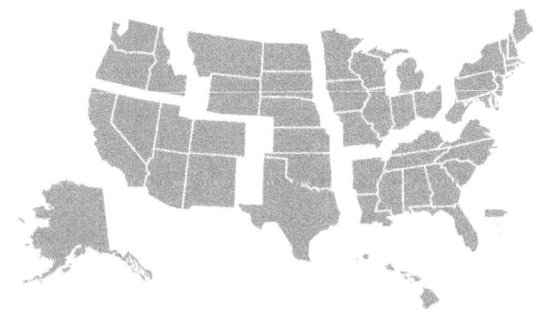

CLIMATE READY
WATER UTILITIES
♻EPA

Climate change in the United States is projected to continue to follow already observable trends. Temperature rise, shifts in precipitation patterns and timing, and altered hydrologic cycles can be expected due to climate change. The following statements, drawn from U.S. Global Change Research Program assessments (USGCRP 2009, USGCRP 2014), are based on projections for climate conditions at the end of the 21st century – using both high and low emissions scenarios (IPCC 2000).

OBSERVED AND PROJECTED CHANGES

- U.S. average temperature has increased by about 1.3 to 1.9°F since 1895, with most of this increase occurring since 1970. In the next few decades, warming is projected to be roughly 2-4°F in most areas. The 2000-2010 decade was the nation's warmest on record.

- Many types of extreme weather events, such as heat waves and regional droughts, have become more frequent and intense during the past 40 to 50 years. Droughts in the Southwest and heat waves everywhere are expected to become more intense in the future.

- Reduced snowpack, reductions in lake ice cover, earlier breakup of ice on lakes and rivers and earlier spring snowmelt have all resulted in earlier peak river flows.

- Cold-season storm tracks are shifting northward due to increasing temperatures, and the strongest storms are likely to become stronger and more frequent.

- The intensity, frequency and duration of North American hurricanes has increased in recent decades, and the intensity of these storms is likely to increase in this century (USGCRP 2014).

GROUP		DW	WW
Drought	Reduced groundwater recharge	●	
	Lower lake & reservoir levels	●	
	Changes in seasonal runoff & loss of snowpack	●	
Water Quality Degradation	Low flow conditions & altered water quality		●
	Saltwater intrusion into aquifers	●	
	Altered surface water quality	●	●
Floods	High flow events & flooding	●	●
	Flooding from coastal storm surges	●	●
Ecosystem Changes	Loss of coastal landforms / wetlands	●	●
	Increased fire risk & altered vegetation	●	●
Service Demand & Use	Volume & temperature challenges	●	●
	Changes in agricultural water demand	●	
	Changes in energy sector needs	●	
	Changes in energy needs of utilities	●	●

Click on a group name above to read more about these impacts or click on a water drop above to read more about a specific impact.

Days Above 100°F in Summer 2011

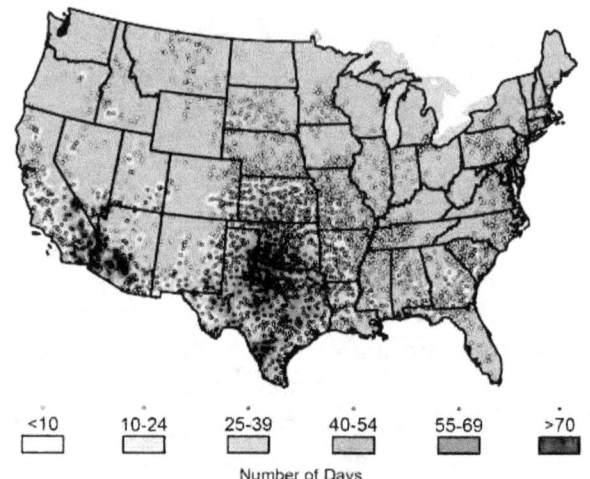

The number of days with a maximum temperature of more than 95°F and number of consecutive hot days is expected to increase. For example, in 2011, cities including Houston, Dallas, Austin, Oklahoma City and Wichita, among others, all set records for the highest number of days recording temperatures of 100°F or higher in those cities' recorded history. In the figure to the right, the circles denote the location of observing stations used in the analysis and the number of recorded 100°F days (NCDC 2012, USGCRP 2014).

<10 10-24 25-39 40-54 55-69 >70

Number of Days

Percentage Change in Very Heavy Precipitation (1958-2012)

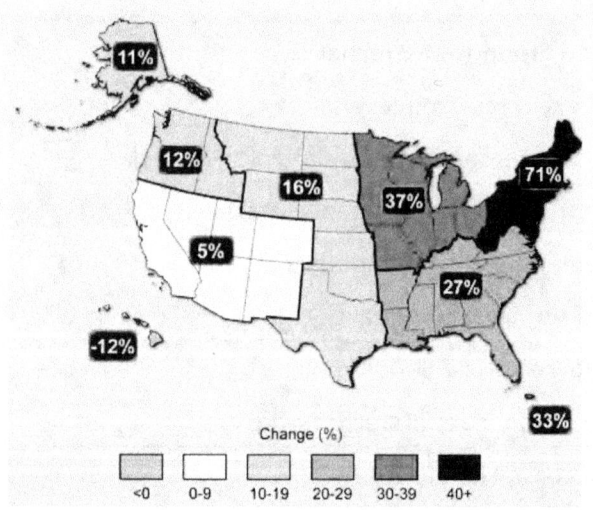

Change (%)

<0 0-9 10-19 20-29 30-39 40+

Throughout the U.S., average annual precipitation has increased by about 5% since 1900. Additionally, the amount of rain falling in the heaviest downpours has increased over the past few decades, with increases of more than 30% in the Northeast, Midwest and Great Plains. This trend is very likely to continue, with the largest increases in the wettest places. The figure to the left shows the percentage increases in the average number of days with very heavy precipitation (defined as the heaviest 1% of all events) from 1958 to 2012 for each region. There are clear trends toward more very heavy precipitation days for the nation as a whole (USGCRP 2014).

Sea level has risen along most of the coast over the last 50 years, and will rise more in the future. Sea level is projected to rise another 1 to 4 feet by 2100, but is not expected to rise uniformly along all coastlines. Regional differences in sea-level rise along U.S. coastlines are illustrated in the map to the right using data from EPA's Climate Resilience Evaluation and Awareness Tool (CREAT). These projections illustrate projected sea-level rise for a scenario with moderate ice melt for 2060. The figure to the right shows that sea level is expected to rise on the Atlantic coast more rapidly than on the Pacific coast. Rising sea level threatens low lying coastal infrastructure and has the potential to exacerbate salt water intrusion of coastal aquifers. Note: These values do not account for subsidence or uplift.

Projected Change in Sea-Level Rise for 2060

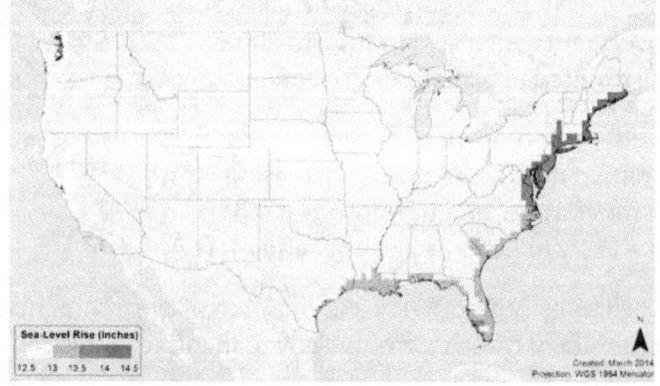

Sea-Level Rise (inches)

12.5 13 13.5 14 14.5

Created: March 2014
Projection: WGS 1984 Mercator

**United States
Environmental Protection
Agency**

Northeast

Climate Region Brief > NORTHEAST

Return to Introduction

Climate change in the northeastern United States is projected to continue to follow already observable trends. Temperature rise, shifts in precipitation patterns and timing, and altered hydrologic cycles can be expected due to climate change. The following statements, drawn from U.S. Global Change Research Program assessments (USGCRP 2009, USGCRP 2014), are based on projections for climate conditions at the end of the 21st century – using both high and low emissions scenarios (IPCC 2000).

OBSERVED AND PROJECTED CHANGES

- Less winter precipitation falling as snow and more as rain is projected. Reduced snowpack, earlier breakup of winter ice on lakes and rivers and earlier spring snowmelt resulting in earlier peak river flows are anticipated.

- Winters in the Northeast are projected to have increased precipitation and be much shorter, with a projected 20-23 fewer days below freezing.

- Short-term droughts (e.g., those lasting from 1 to 3 months) are projected to occur as frequently as once each summer in the Catskill and Adirondack Mountains and across the New England states.

- Sea level in this region is projected to rise at a rate greater than the global average, between 1 to 4 feet by 2100. Severe flooding due to sea-level rise and heavy downpours are likely to occur more frequently (Parris et al. 2012).

GROUP		DW	WW
Drought	Reduced groundwater recharge	💧	
	Lower lake & reservoir levels	💧	
	Changes in seasonal runoff & loss of snowpack	💧💧	
Water Quality Degradation	Low flow conditions & altered water quality		💧💧
	Saltwater intrusion into aquifers	💧	
	Altered surface water quality	💧	💧
Floods	High flow events & flooding	💧💧	💧💧
	Flooding from coastal storm surges	💧💧	💧💧
Ecosystem Changes	Loss of coastal landforms / wetlands	💧💧	💧💧
	Increased fire risk & altered vegetation	💧	💧
Service Demand & Use	Volume & temperature challenges	💧💧	💧💧
	Changes in agricultural water demand	💧	
	Changes in energy sector needs	💧	
	Changes in energy needs of utilities	💧💧	💧💧

Click on a group name above to read more about these impacts or click on a water drop above to read more about a specific impact.

💧💧 = Particularly relevant to Northeast   💧 = Somewhat relevant

- Increases in the extent and frequency of storm surge and coastal flooding would increase erosion, property damage and loss of wetlands.

Continued on page 2

Projected Increase in the Number of Days over 90°F

Much of the southern portion of the region is projected to experience more than 60 additional days per year above 90°F by the 2050s. The figure to the right shows model projections of the increased number of summer days with temperatures above 90°F between 2041-2070, compared to 1970-2000, using a higher emissions scenario (NOAA NCDC/CICS-NC, USGCRP 2014).

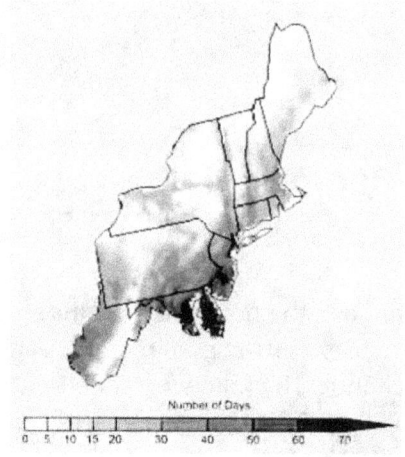

Projected Change in Intense Precipitation (1-in-100 Year Storm) for 2060

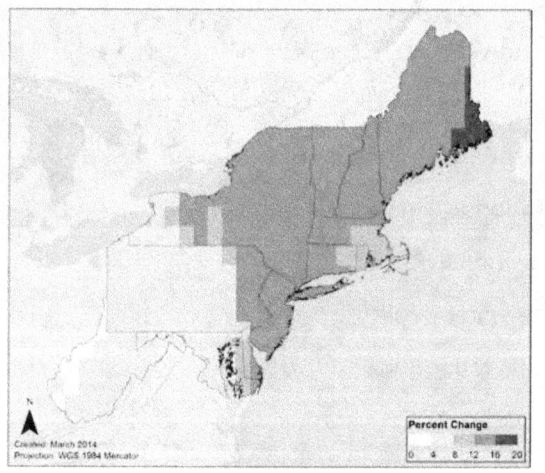

The amount of rain falling in the heaviest downpours has been increasing, and this trend is projected to continue. Between 1958 and 2010, the Northeast saw more than a 70% increase in the average number of days with very heavy precipitation (Groisman et al. 2013). Projected changes in intense precipitation in the Northeast could have significant impacts for drinking water and wastewater utilities (e.g., facility inundation and resulting infrastructure damage, increase in combined sewer overflows, increased pollutant and sediment loading). The figure to the left shows projected changes in the magnitude of the 1-in-100 year storm from current conditions using data from EPA's Climate Resilience Evaluation and Awareness Tool (CREAT) for 2060. In most of the Northeast, the magnitude of these events is projected to increase anywhere from 4 to 20%.

Simulated Change in Seasonal Mean Precipitation
(A2 Scenario, 2041-2070 minus 1980-2000)

Annual mean precipitation is projected to increase; however, seasonal differences are expected. As shown in the figure to the right, seasonal projections indicate that average precipitation will increase in the winter and spring for the entire Northeast, will increase for most of the Northeast in the fall and will decrease in general during the summer. The projections shown are for 2041-2070 under a higher emissions scenario (A2) (Kunkel et al. 2013).

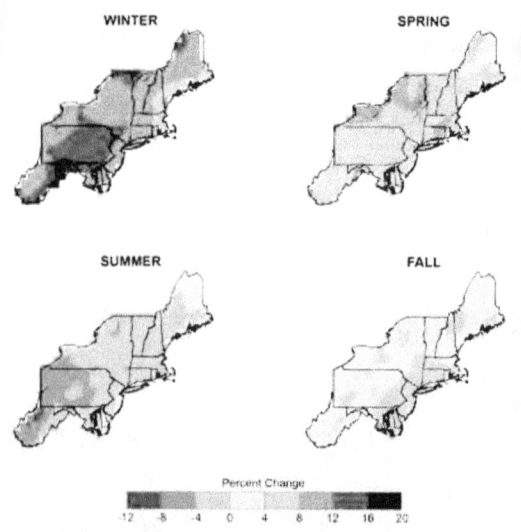

CLIMATE READY
WATER UTILITIES
♻EPA

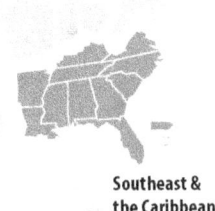

**Southeast &
the Caribbean**

Climate Region Brief > SOUTHEAST & THE CARIBBEAN

Return to Introduction

Climate change in the southeastern United States and the Caribbean is projected to continue to follow already observable trends. Temperature rise, shifts in precipitation patterns and timing and altered hydrologic cycles can be expected due to climate change. The following statements, drawn from U.S. Global Change Research Program assessments (USGCRP 2009, USGCRP 2014), are based on projections for climate conditions at the end of the 21st century – using both high and low emissions scenarios (IPCC 2000).

OBSERVED AND PROJECTED CHANGES

- Average annual temperatures are expected to increase between 4 to 8°F in the region by 2100, with projected temperature increases for the interior states of this region being 1 to 2°F higher than the coastal regions. Projected temperature increases for Puerto Rico are 2 to 5°F by 2100 (Kunkel et al. 2013).

- The frequency, duration and intensity of droughts are likely to continue to increase, leading to the drying up of lakes, ponds and wetlands; reduced groundwater recharge; and an increased risk of flash flooding if the dry ground cannot effectively absorb rainwater.

- Dissolved oxygen in streams, lakes and shallow aquatic habitats is likely to decline.

- Historical weather records indicate that the frequency of extreme precipitation events has been increasing across the Southeast. This observed trend is expected to continue. Summers are expected to trend toward extremes, being either exceptionally wet or exceptionally dry (Kunkel et al. 2013).

- Fewer tropical storms are projected, but these storms are expected to be of greater intensity – with more Category 4 and 5 events. Higher peak wind speeds, rainfall intensity and storm surge height and strength would also increase inland and coastal flooding, coastal erosion rates, wind damage to coastal forests and wetland loss (Knutson et al. 2013).

- More frequent storm surge flooding and permanent inundation of coastal ecosystems and communities is likely in some low-lying areas, particularly along the central Gulf Coast where the land surface is sinking due to geological tectonics, consolidation of sediment and groundwater pumping.

- Sea level will gradually rise to a critical elevation, resulting in rapid saltwater intrusion into freshwater aquifers, and salinity increases in estuaries, coastal wetlands and tidal rivers (SFWMD 2009, Obeysekera et al. 2011).

GROUP		DW	WW
Drought	Reduced groundwater recharge	💧💧	
	Lower lake & reservoir levels	💧💧	
	Changes in seasonal runoff & loss of snowpack	💧	
Water Quality Degradation	Low flow conditions & altered water quality		💧
	Saltwater intrusion into aquifers	💧💧	
	Altered surface water quality	💧💧	💧💧
Floods	High flow events & flooding	💧	💧
	Flooding from coastal storm surges	💧💧	💧💧
Ecosystem Changes	Loss of coastal landforms / wetlands	💧💧	💧💧
	Increased fire risk & altered vegetation	💧💧	💧💧
Service Demand & Use	Volume & temperature challenges	💧	💧
	Changes in agricultural water demand	💧💧	
	Changes in energy sector needs	💧	
	Changes in energy needs of utilities	💧💧	💧💧

Click on a group name above to read more about these impacts or click on a water drop above to read more about a specific impact.

💧💧 = Particularly relevant to Southeast and the Caribbean   💧 = Somewhat relevant

---

Continued on page 2

Adaptation Strategies Guide for Water Utilities
TABLE OF CONTENTS

CLIMATE REGION BRIEFS

National
Northeast
Southeast
Midwest
Great Plains

Southwest
Northwest
Alaska
Pacific Islands
Coasts

STRATEGY BRIEFS

 Group: Drought
Reduced Groundwater Recharge (DW)
Lower Lake & Reservoir Levels (DW)
Changes in Seasonal Runoff & Loss of Snowpack (DW)

 Group: Water Quality Degradation
Low Flow Conditions & Altered Water Quality (WW)
Saltwater Intrusion into Aquifers (DW)
Altered Surface Water Quality (DW)
Altered Surface Water Quality (WW)

 Group: Floods
High Flow Events & Flooding (DW)
High Flow Events & Flooding (WW)
Flooding from Coastal Storm Surges (DW)
Flooding from Coastal Storm Surges (WW)

 Group: Ecosystem Changes
Loss of Coastal Landforms / Wetlands (DW/WW)
Increased Fire Risk & Altered Vegetation (DW/WW)

 Group: Service Demand and Use
Volume & Temperature Challenges (DW)
Volume & Temperature Challenges (WW)
Changes in Agricultural Water Demand (DW)
Changes in Energy Sector Needs & Energy Needs of Utilities (DW/WW)

SUSTAINABILITY BRIEFS

 Green Infrastructure

 Energy Management

 Water Demand Management

Billion Dollar Weather/Climate Disasters

Many types of extreme weather events, such as heat waves and regional droughts, have become more frequent and intense during the past 40 to 50 years. The figure to the right shows the number of weather/climate disasters from 1980-2012 that resulted in more than $1 billion in damages. The Southeast has experienced more billion-dollar disasters than any other region (NOAA 2013).

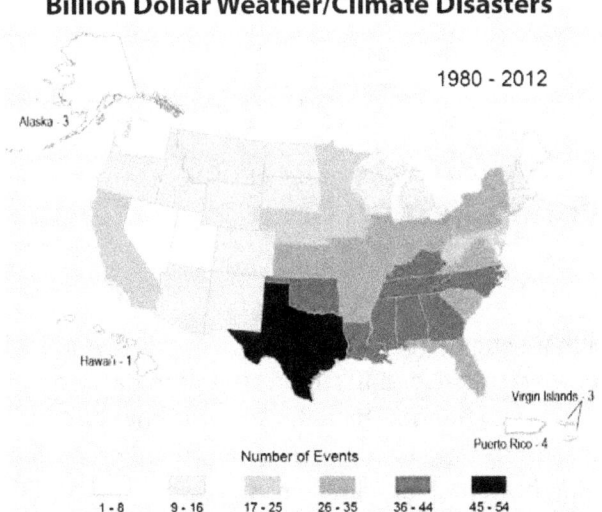

1980 - 2012

Number of Events

| 1 - 8 | 9 - 16 | 17 - 25 | 26 - 35 | 36 - 44 | 45 - 54 |

Trends in Water Availability

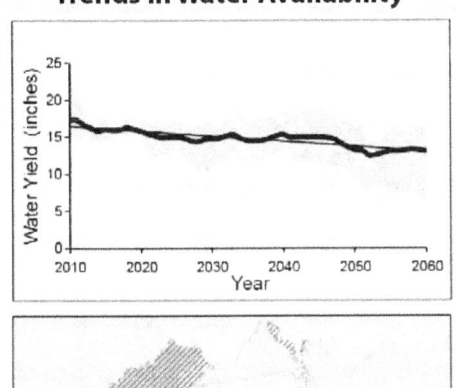

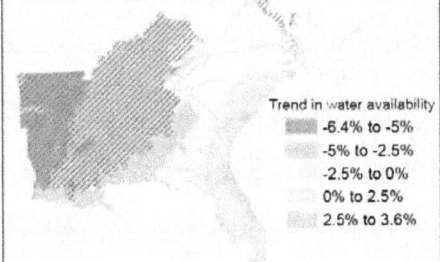

Trend in water availability
- -6.4% to -5%
- -5% to -2.5%
- -2.5% to 0%
- 0% to 2.5%
- 2.5% to 3.6%

Net water supply in the Southeast is expected to decline over the next several decades due to droughts, increased demand due to higher temperatures and competing uses (e.g., agriculture). The top panel shows projected 10-year moving average annual water yield based on a mid-range and low emissions scenario. The bottom panel shows average annual water yield (equivalent to water availability) trends projected for 2010-2060 (under mid-range and low emissions scenarios) compared to the average from 2001-2010. The western part of the Southeast is expected to see the largest reductions in water availability. Statistical confidence in the data is highest in the hatched areas (Sun et al. 2013). Analysis of current and future water resources in the Caribbean shows many of the small islands would be exposed to severe water stress under all climate scenarios (UNEP 2008).

Projected Change in Average Annual Precipitation for 2060

In general, annual precipitation is projected to decrease across the region (USGCRP 2014). The figure to the right shows projected changes in annual precipitation for 2060 using data from EPA's Climate Resilience Evaluation and Awareness Tool (CREAT). Data shows that decreases in precipitation are projected for the region, with the largest percent decrease in Florida. These changes could exacerbate existing challenges related to surface water quality that drinking water and wastewater utilities already face.

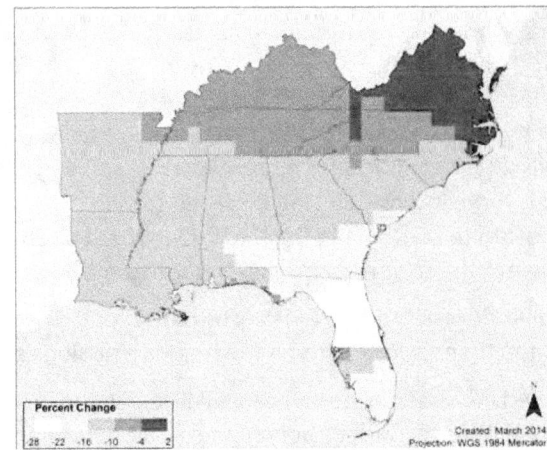

Percent Change

| -28 | -22 | -16 | -10 | -4 | 2 |

Created: March 2014
Projection: WGS 1984 Mercator

United States
Environmental Protection
Agency

Midwest

CLIMATE READY
WATER UTILITIES
♻EPA

Climate Region Brief > MIDWEST

Return to Introduction

Climate change in the midwestern United States is projected to continue to follow already observable trends. Temperature rise, shifts in precipitation patterns and timing and altered hydrologic cycles can be expected due to climate change. The following statements, drawn from U.S. Global Change Research Program assessments (USGCRP 2009, USGCRP 2014), are based on projections for climate conditions at the end of the 21st century – using both high and low emissions scenarios (IPCC 2000).

OBSERVED AND PROJECTED CHANGES

- Heat waves are anticipated to be more frequent, more severe and longer in duration.

- As air temperatures increase, so will surface water temperatures and frequency of algal blooms. In some lakes, mixing of the warmer surface lake water with the colder water below will be reduced; this stratification can cut off oxygen from bottom layers, increasing the risk of oxygen-poor or oxygen-free "dead zones" (Reutter et al. 2011).

- In lakes with contaminated sediment, warmer water and low-oxygen conditions can more readily release mercury and other persistent pollutants into surface water.

- Reduced summer water levels are also likely to reduce the recharge of groundwater, dry up small streams and reduce the area of wetlands in the Midwest.

- Generally, annual precipitation increased during the past century (by up to 20% in some locations), with much of the increase driven by intensification of the heaviest rainfalls (Pryor et al. 2009). This tendency towards more intense precipitation events

GROUP		DW	WW
Drought	Reduced groundwater recharge	💧💧	
Drought	Lower lake & reservoir levels	💧💧	
Drought	Changes in seasonal runoff & loss of snowpack	💧	
Water Quality Degradation	Low flow conditions & altered water quality		💧
Water Quality Degradation	Saltwater intrusion into aquifers	💧	
Water Quality Degradation	Altered surface water quality	💧💧	💧💧
Floods	High flow events & flooding	💧💧	💧💧
Floods	Flooding from coastal storm surges	💧	💧
Ecosystem Changes	Loss of coastal landforms / wetlands	💧	💧
Ecosystem Changes	Increased fire risk & altered vegetation	💧	💧
Service Demand & Use	Volume & temperature challenges	💧💧	💧💧
Service Demand & Use	Changes in agricultural water demand	💧💧	
Service Demand & Use	Changes in energy sector needs	💧💧	
Service Demand & Use	Changes in energy needs of utilities	💧💧	💧💧

Click on a group name above to read more about these impacts or click on a water drop above to read more about a specific impact.

💧💧 = Particularly relevant to Midwest   💧 = Somewhat relevant

is projected to continue in the future (Schoof et al. 2010). Precipitation is projected to increase in winter, spring and fall, but decrease in the summer, and the average number of days each year without precipitation is expected to increase.

- Rainfall-induced flooding is projected to occur twice as often by the end of this century under the lower emissions scenario, and three times as often under the higher emissions scenario.

- Projected increases in storm events will lead to an increase of up to 120% in Combined Sewer Overflows (CSOs) into Lake Michigan by 2100 under a very high emissions scenario (Patz et al. 2008).

Continued on page 2

The Great Lakes have recently recorded higher water temperatures and less ice cover. Due to reduction in its ice cover, the temperatures of surface water in Lake Superior increased 4.5°F in the summer, twice the rate of increase in air temperature during summers between 1968 and 2002 (Austin et al. 2007). Great Lake surface temperatures are projected to rise by as much as 7°F by 2050 and 12.1°F by 2100 (Mackey et al. 2012, Trumpickas et al. 2009). The top panel of the figure to the right shows the decline in average annual percentage of the Great Lakes covered with ice from 1970 to 2010. Photos in the bottom panel contrast extensive vs. minimal ice cover on Lake Erie (Wang et al. 2012). Winter of 2008-2009 (lower left) was characterized by near-normal air temperatures over the Great Lakes, while 2011-2012 (lower right) was characterized by air temperature of approximately 5.4°F warmer than the historical average (images sourced from NASA MODIS satellite imagery processed by SSEC, University of Wisconsin and obtained from CoastWatch Great Lakes Program).

Declining Ice Cover on the Great Lakes

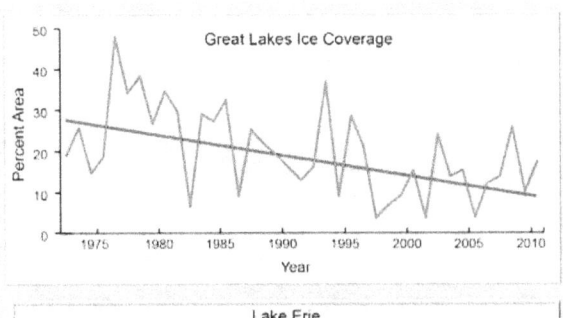

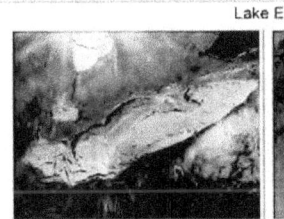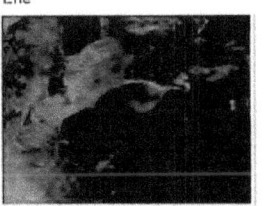

Annual precipitation has increased in the past century (by up to 20% in some locations), with much of the increase coming from heavy rain events. The four maps below show projected changes (for 2041-2070 [relative to 1971-2000]) in (a) annual average precipitation, (b) heavy precipitation (top 2% of all rainfalls), (c) the increases in the amount of rain falling in the wettest 5-day period and (d) change in the average number of consecutive dry days with less than 0.01 inches of precipitation. An increase in consecutive dry days has been used to indicate an increase in future droughts (NOAA NCDC/CICS-NC, IPCC, USGCRP 2014).

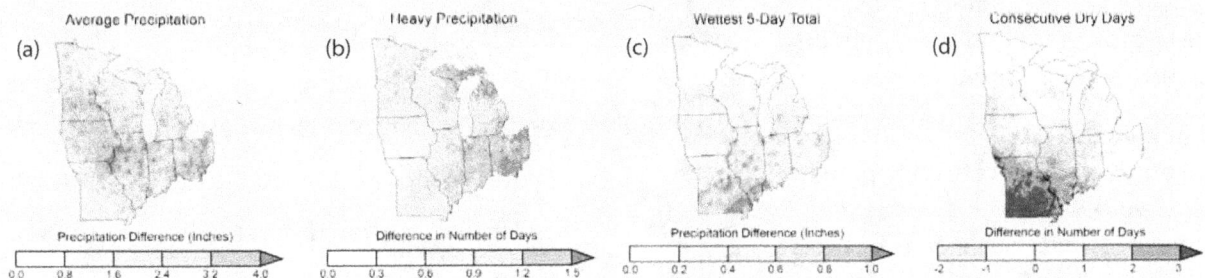

Average annual temperature is projected to increase in the Midwest. The figure to the right shows projected changes in annual temperature for 2060 using data from EPA's Climate Resilience Evaluation and Awareness Tool (CREAT). In most of the Midwest, the average annual temperature is projected to increase by about 5 to 6° F under a hot/dry scenario compared to current conditions, with the largest temperature increases projected for northern Minnesota. Projected changes in seasonal or monthly temperature may be even more dramatic, causing disruptions to hydrologic cycles and management of drinking water and wastewater utilities.

Projected Change in Temperature for 2060

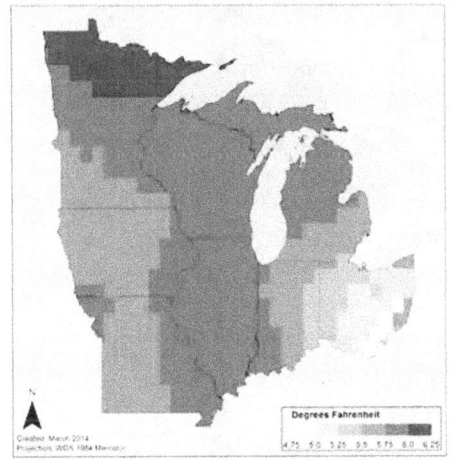

 **EPA** United States
Environmental Protection
Agency

CLIMATE READY
WATER UTILITIES
♻EPA

Great Plains

Climate change in the Great Plains of the United States is projected to continue to follow already observable trends. Temperature rise, shifts in precipitation patterns and timing and altered hydrologic cycles can be expected due to climate change. The following statements, drawn from U.S. Global Change Research Program assessments (USGCRP 2009, USGCRP 2014), are based on projections for climate conditions at the end of the 21st century – using both high and low emissions scenarios (IPCC 2000).

OBSERVED AND PROJECTED CHANGES

- Projections of increasing temperatures, faster evaporation rates and more sustained droughts brought on by climate change will only add more stress to overtaxed water sources.

- Further stresses on agricultural water supply are likely as the region's cities continue to grow, increasing competition between urban and rural water users.

- Precipitation is also projected to change in all seasons. Conditions are anticipated to become wetter in the north and drier in the south. However, large parts of Texas and Oklahoma are projected to have an increase in the number of days with no precipitation (an increase of up to 5 or more days per year by mid-century).

- Rapid spring warming and intense rainfall could increase runoff and cause flooding, reducing water quality and eroding soils.

- Projected increases in precipitation are unlikely to be sufficient to offset decreasing soil moisture and water availability in the Great Plains, due to rising temperatures and aquifer depletion.

- More frequent extreme events, such as heat waves, droughts, snow and heavy rainfall are projected to occur.

GROUP		DW	WW
Drought	Reduced groundwater recharge	💧💧	
	Lower lake & reservoir levels	💧💧	
	Changes in seasonal runoff & loss of snowpack	💧	
Water Quality Degradation	Low flow conditions & altered water quality		💧
	Saltwater intrusion into aquifers	💧	
	Altered surface water quality	💧	💧
Floods	High flow events & flooding	💧💧	💧💧
	Flooding from coastal storm surges	💧	💧
Ecosystem Changes	Loss of coastal landforms / wetlands	💧	💧
	Increased fire risk & altered vegetation	💧💧	💧💧
Service Demand & Use	Volume & temperature challenges	💧💧	💧💧
	Changes in agricultural water demand	💧💧	
	Changes in energy sector needs	💧	
	Changes in energy needs of utilities	💧💧	💧💧

Click on a group name above to read more about these impacts or click on a water drop above to read more about a specific impact.

💧💧 = Particularly relevant to Great Plains   💧 = Somewhat relevant

Temperatures are projected to continue to increase over this century, with summer increases larger than winter increases in the southern and central Great Plains. This figure shows changes in the mean annual freeze-free season length for the Great Plains. The freeze-free season length has generally been increasing since the early 20th century. The last occurrence of 32°F in the spring has been occurring earlier, and the first occurrence of 32°F in the fall has been happening later (Kunkel et al. 2013).

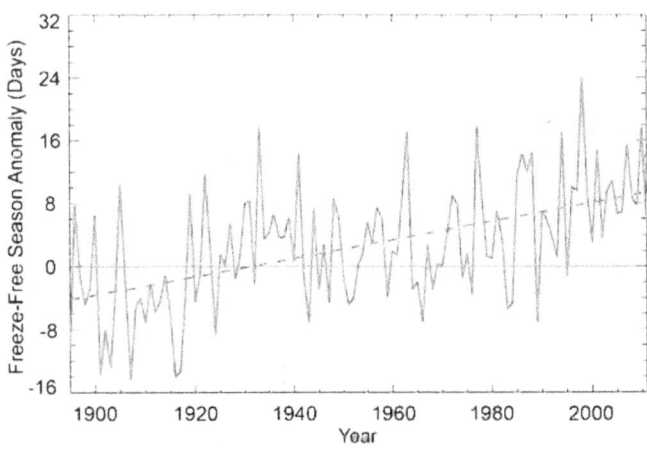

Difference in Mean Annual Freeze-Free Season Length for the U.S. Great Plains
(Deviations from the 1901-1960 Average)

Projected Change in Number of Dry Days

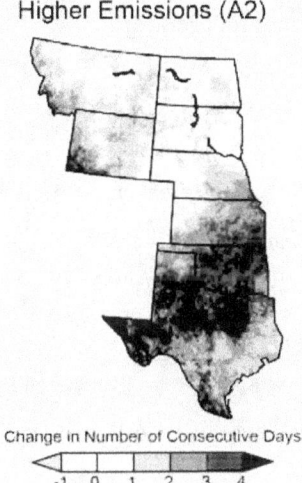

Higher Emissions (A2)

Change in Number of Consecutive Days

Current trends in the Great Plains of a drier south and a wetter north are projected to become more pronounced by 2050. The figure to the left shows the projected change in number of days with less than 0.01 inches of precipitation under a higher emissions scenario (IPCC 2000) for 2041-2070, compared to 1971-2000 averages. The southeastern Great Plains, which is the wettest portion of the region, is projected to experience large increases in the number of consecutive dry days (NOAA NCDC/CIC-NC, USGCRP 2014).

The number of days with heavy precipitation (>1 inch) is expected to increase by mid-century, especially in the northern Great Plains. The figure to the right shows projected changes in the magnitude of the 1 in 100 year storm from current conditions using data from EPA's Climate Resilience Evaluation and Awareness Tool (CREAT) for 2060. The magnitude of the 100-year storm is projected to increase from 5-15% for most of the Great Plains. In parts of the northern and mid-Great Plains and the Texas Gulf Coast, the magnitude of the 100-year storm is projected to increase by more than 15%. Increases in the magnitude of heavy precipitation events will result in greater challenges to drinking water and wastewater utilities (e.g., facility inundation and resulting infrastructure damage, increased combined sewer overflows and increased pollutant and sediment loading).

Projected Change in Intense Precipitation (1-in-100 Year Storm) for 2060

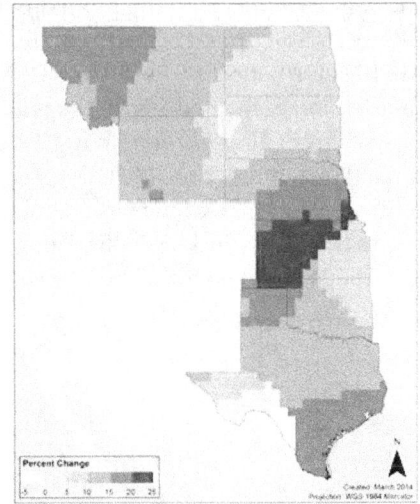

Percent Change

 United States Environmental Protection Agency

CLIMATE READY
WATER UTILITIES
⚡EPA

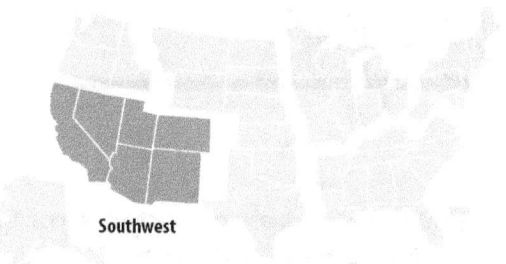

Southwest

Climate Region Brief > SOUTHWEST

Return to Introduction

Climate change in the southwestern United States is projected to continue to follow already observable trends. Temperature rise, shifts in precipitation patterns and timing and altered hydrologic cycles can be expected due to climate change. The following statements, drawn from U.S. Global Change Research Program assessments (USGCRP 2009, USGCRP 2014), are based on projections for climate conditions at the end of the 21st century – using both high and low emissions scenarios (IPCC 2000).

OBSERVED AND PROJECTED CHANGES

- The 2001-2010 decade was the warmest on record. Average observed temperatures in the Southwest were almost 2°F higher than historic averages, with the region experiencing more heat waves and fewer cold snaps.

- Projected increases in summertime temperatures are greater than the increase of annual average temperature in parts of the region and will likely be exacerbated locally by expanding urban heat island effects.

- Less winter precipitation falling as snow and earlier spring snow melt are projected to shift runoff and most of the annual streamflow to earlier in the year.

- Future droughts are projected to be substantially hotter. For major river basins, such as the Colorado River Basin, drought is projected to become more frequent, intense and longer lasting than in the historical record (Cayan et al. 2012).

- Increasing temperature will cause more droughts, wildfires and invasive species colonization, which will accelerate transformation of the landscape. Models project a doubling of burned area in the Southern Rockies (Litschert et al. 2012) and up to 74% more fires in California (Westerling et al. 2012). The area burned in the Southwest has increased by more than 300% compared to the 1970s and 1980s. Drought has been widespread in the Southwest since 2000; the drought conditions during the 2000s were the most severe average drought conditions of any decade.

- Increased flood risk in the Southwest is likely to result from a combination of decreased snow cover on the lower slopes of high mountains and an increased fraction of winter precipitation falling as rain, which will run off more rapidly and alter the timing of flooding.

GROUP		DW	WW
Drought	Reduced groundwater recharge	◔	
	Lower lake & reservoir levels	◔	
	Changes in seasonal runoff & loss of snowpack	◔◔	
Water Quality Degradation	Low flow conditions & altered water quality		◔◔
	Saltwater intrusion into aquifers	◔	
	Altered surface water quality	◔	◔
Floods	High flow events & flooding	◔◔	◔◔
	Flooding from coastal storm surges	◔	◔
Ecosystem Changes	Loss of coastal landforms / wetlands	◔	◔
	Increased fire risk & altered vegetation	◔◔	◔◔
Service Demand & Use	Volume & temperature challenges	◔◔	◔◔
	Changes in agricultural water demand	◔◔	
	Changes in energy sector needs	◔◔	
	Changes in energy needs of utilities	◔◔	◔◔

Click on a group name above to read more about these impacts or click on a water drop above to read more about a specific impact.

◔◔ = Particularly relevant to Southwest   ◔ = Somewhat relevant

Continued on page 2

Reduced water content of snowpack or snow water equivalent, runoff and soil moisture are projected in the Southwest. These figures show percentage projected changes for mid-century (2041 to 2070) under a higher emissions scenario (IPCC 2000). These projections illustrate: (a) major losses in the water content of snowpack that fills western rivers; (b) significant reductions in runoff in California, Arizona and the Central Rocky Mountains and (c) reductions in soil moisture across the Southwest. Decline in snowpack and streamflow and more frequent dry winters suggest an increased risk for systems to experience water shortages (Cayan et al. 2013, USGCRP 2014).

Changes in Snow, Runoff and Soil Moisture

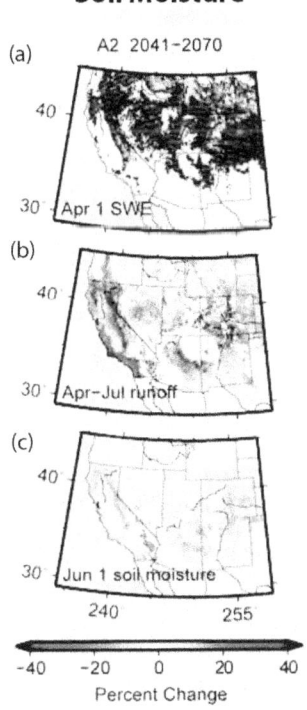

The frequency of heat waves has generally been increasing in recent decades, with a statistically significant upward trend. More intense, longer-lasting heat wave events are projected to occur over this century. There is an overall downward trend in the occurrence of cold waves that is also statistically significant. The graphs below show the changes from 1900-2000 in the heat wave (top panel) and cold wave (bottom panel) index for the Southwest (Kunkel et al. 2013).

Mean Annual Heat Wave (top) and Cold Wave (bottom) Index for the Southwest U.S.
(Occurance of 4-day, 1 in 5-year events)

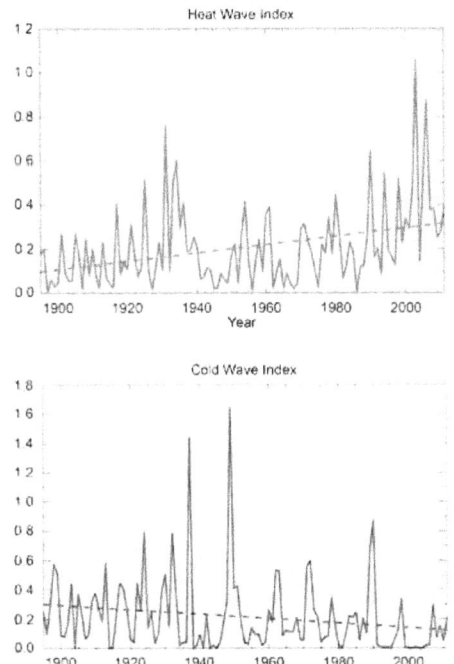

Average annual precipitation is projected to decrease throughout the Southwest. The figure to the right shows projected changes in annual precipitation for 2060 using data from EPA's Climate Resilience Evaluation and Awareness Tool (CREAT). Conditions are projected to be drier for the entire region, and in some portions of the Southwest, average annual precipitation is projected to decrease by 25-30% compared to current conditions. Increases in the annual maximum number of consecutive dry days, up to 26 days above present-day values, are expected for parts of southern California and Arizona. These precipitation projections contribute to the increased probability of more severe droughts for the region.

Projected Change in Average Annual Precipitation for 2060

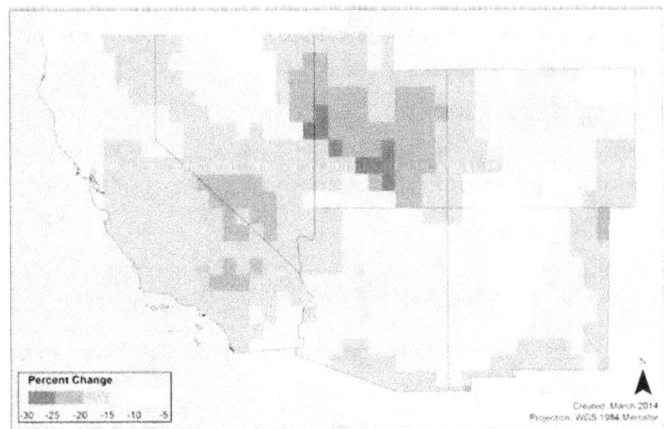

United States Environmental Protection Agency

CLIMATE READY
WATER UTILITIES

Northwest

Climate change in the northwestern United States is projected to continue to follow already observable trends. Temperature rise, shifts in precipitation patterns and timing and altered hydrologic cycles can be expected due to climate change. The following statements, drawn from U.S. Global Change Research Program assessments (USGCRP 2009, USGCRP 2014), are based on projections for climate conditions at the end of the 21st century – using both high and low emissions scenarios (IPCC 2000).

OBSERVED AND PROJECTED CHANGES

- Average annual temperature for the Northwest is projected to increase by about 3 to 10°F during this century. The number of hot days (maximum temperatures over 95°F) is projected to increase, with the largest increase in the southeastern part of the region.

- By 2050, snowmelt is projected to shift 3 to 4 weeks earlier than the 20th century average. Areas dominated by rain, rather than snow, are not expected to see major shifts in the timing of runoff.

- April 1 snowpack has declined 20% since the 1950s, a trend that is projected to continue, leading to earlier peak streamflow and a reduction in the amount of water available during the warm season.

- Increasing winter rainfall (as opposed to snowfall) is expected to lead to more winter flooding and an increased number of landslides due to saturated soils.

- Sensitive watersheds are projected to experience both increased flood risk in winter and increased drought risk in summer due to warming.

GROUP		DW	WW
Drought	Reduced groundwater recharge	💧	
	Lower lake & reservoir levels	💧	
	Changes in seasonal runoff & loss of snowpack	💧💧	
Water Quality Degradation	Low flow conditions & altered water quality		💧💧
	Saltwater intrusion into aquifers	💧	
	Altered surface water quality	💧	💧
Floods	High flow events & flooding	💧💧	💧💧
	Flooding from coastal storm surges	💧	💧
Ecosystem Changes	Loss of coastal landforms / wetlands	💧💧	💧💧
	Increased fire risk & altered vegetation	💧💧	💧💧
Service Demand & Use	Volume & temperature challenges	💧💧	💧💧
	Changes in agricultural water demand	💧	
	Changes in energy sector needs	💧	
	Changes in energy needs of utilities	💧💧	💧💧

Click on a group name above to read more about these impacts or click on a water drop above to read more about a specific impact.

💧💧 = Particularly relevant to Northwest   💧 = Somewhat relevant

- Precipitation is expected to increase in all seasons except the summer. Drier summers will lead to a greater risk of wildfires throughout the region.

- Low streamflows in late summer are projected to be even lower due to drought and reduced summer precipitation (as much as 30% reduction by 2100).

- Sea-level rise will increase erosion of the Northwest coast and cause the loss of beaches and other significant coastal land.

Low streamflows in late summer are projected to be even lower due to drought and reduced summer precipitation. Decreases in summer flows can lead to water shortages. The figure to the right shows June streamflow trends from 1948 to 2008. The fraction of annual flow arriving in June in snow-fed rivers increased slightly in rain-dominated coastal basins and decreased in mixed rain-snow basins and snowmelt-dominated basins for 1948-2008 (Fritze et al. 2011).

Observed Shifts in Streamflow Timing

June Streamflow Trends
(fraction of annual flow)
1948-2008
● -15% to -8%
● -8% to -4%
◑ -4% to -2%
○ -2% to -1%
○ -1% to 0%
○ 0% to +1%
○ +1% to +2%
○ +2% to +3%
Elevation
< 300 ft
300 ft - 1500 ft
1500 ft - 3000 ft
3000 ft - 6000 ft
> 6000 ft

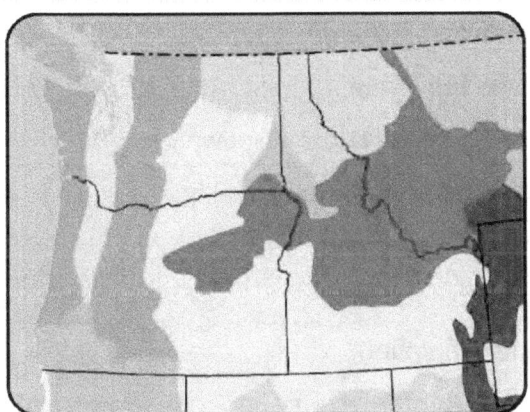

Projected Increase
in Area Burned
600% to 700%
500% to 600%
400% to 500%
300% to 400%
200% to 300%
100% to 200%
Not modeled

Large increases in area burned by wildfire are projected for most of the Northwest. The figure to the left shows the projected increase in area burned, considering temperature and precipitation changes associated with a 2.2°F global warming (NRC 2011). The divisions of the different shaded locations are areas that share common climatic and vegetation characteristics (Bailey 1995).

Projected Change in Intense Precipitation (1-in-100 Year Storm) for 2060

An increase in extreme daily precipitation is projected for the Northwest, with a 13% increase in number of days with more than 1 inch of precipitation. The figure to the right shows projected changes in the magnitude of 1-in-100 year storm from current conditions using data from EPA's Climate Resilience Evaluation and Awareness Tool (CREAT) for 2060. The magnitude of the 100-year storm is projected to increase about 4% along most of the Pacific coast, 8 to 12% in the central portion of the region and 12 to 20% in the southeastern portion of the region.

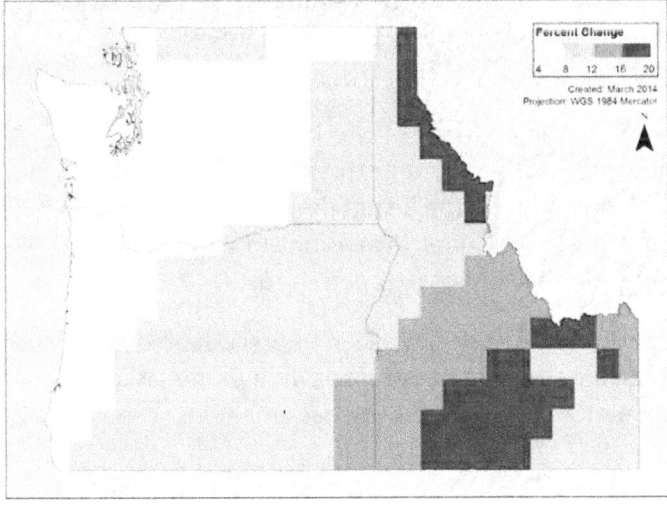

Percent Change
4   8   12   16   20
Created: March 2014
Projection: WGS 1984 Mercator

CLIMATE READY
WATER UTILITIES
♻EPA

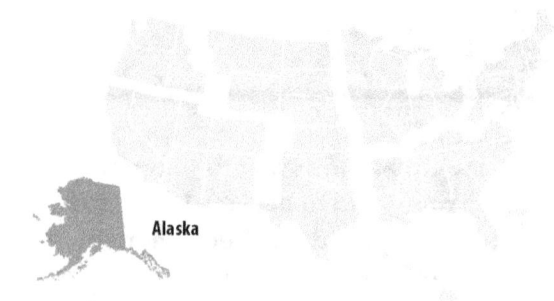
Alaska

Climate Region Brief > ALASKA

Return to Introduction

Climate change in Alaska is projected to continue to follow already observable trends. Temperature rise, shifts in precipitation patterns and timing and altered hydrologic cycles can be expected due to climate change. The following statements, drawn from U.S. Global Change Research Program assessments (USGCRP 2009, USGCRP 2014), are based on projections for climate conditions at the end of the 21st century – using both high and low emissions scenarios (IPCC 2000).

OBSERVED AND PROJECTED CHANGES

- Average annual temperatures in this region are projected to rise about 2 to 4°F by the middle of this century under a lower emissions scenario. However, by the end of the century, northern parts of the region are projected to warm by 6 to 8°F, while the rest of the region is projected to warm by 4 to 6°F.

- Higher air temperatures will increase evaporation rates, reducing water availability and storage. In addition, these increasing temperatures will continue to reduce Arctic sea ice coverage.

- Annual precipitation is projected to increase by an average of 25% under a high emissions scenario by the end of the century.

- Alaska's coastlines, many of which are low in elevation, are increasingly threatened by a combination of the loss of their protective sea-ice buffer, increasing storm activity and thawing coastal permafrost.

GROUP		DW	WW
Drought	Reduced groundwater recharge	◔	
	Lower lake & reservoir levels	◔	
	Changes in seasonal runoff & loss of snowpack	◔	
Water Quality Degradation	Low flow conditions & altered water quality		◔
	Saltwater intrusion into aquifers	◔	
	Altered surface water quality	◔	◔
Floods	High flow events & flooding	◔	◔
	Flooding from coastal storm surges	◔	◔
Ecosystem Changes	Loss of coastal landforms / wetlands	◔◔	◔◔
	Increased fire risk & altered vegetation	◔◔	◔◔
Service Demand & Use	Volume & temperature challenges	◔	◔
	Changes in agricultural water demand	◔	
	Changes in energy sector needs	◔◔	
	Changes in energy needs of utilities	◔◔	◔◔

Click on a group name above to read more about these impacts or click on a water drop above to read more about a specific impact.

◔◔ = Particularly relevant to Alaska   ◔ = Somewhat relevant

- The past several years have seen unprecedented fire occurrences on the tundra of northern and western Alaska. The average area burned per year by wildfires in Alaska is projected to double by the middle of this century and triple under a moderate greenhouse gas emissions scenario by 2100.

This page left intentionally blank

Average annual temperatures in Alaska are projected to increase. Northern latitudes are warming faster than more temperate regions, and Alaska has already warmed much faster than the rest of the country. The figure to the right shows projected changes in temperature, relative to 1971-1999, projected for Alaska in the early, middle and late parts of this century. Projections are provided for both higher and lower emissions scenarios (Stewart et al. 2013).

Projected Change in Average Annual Temperature for Two Emissions Scenarios

2021–2050   2041–2070   2070–2099

Lower Emissions (B1)

Temperature Change (°F)
1.5 3.5 5.5 7.5 9.5 11.5 13.5

Projected Average Soil Temperature for Two Emissions Scenarios

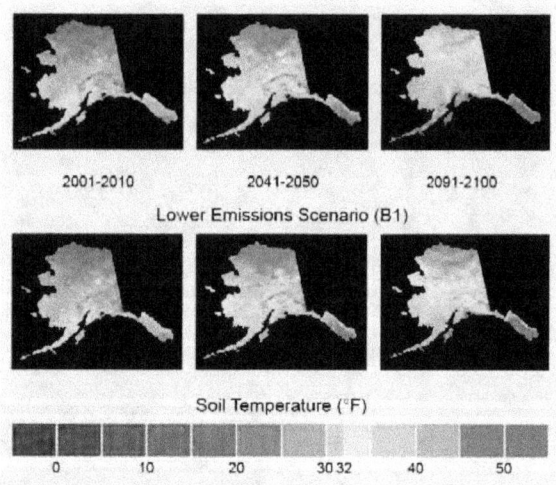

2001-2010   2041-2050   2091-2100

Lower Emissions Scenario (B1)

Soil Temperature (°F)
0   10   20   30 32   40   50

Increasing temperatures will result in thawing permafrost, the extent of which can be seen from soil temperatures. The figure to the left shows the simulated annual mean soil temperature at 1-meter depth under high and low emissions scenarios. Permafrost degradation is projected to increase for the mid and late 21st century for both scenarios Thawing permafrost can mobilize subsurface water, reroute surface water and damage roads, runways, water and sewer systems and other infrastructure (Permafrost Lab, Geophysical Institute, University of Alaska Fairbanks).

Annual precipitation is projected to increase throughout Alaska. The figure to the right shows projected changes in average annual precipitation for 2060 using data from EPA's Climate Resilience Evaluation and Awareness Tool (CREAT). Data shows that a 10 to 25% increase in average annual precipitation is projected for the entire state, which could exacerbate challenges such as coastal erosion and stormwater management that drinking water and wastewater utilities already face.

Projected Change in Average Annual Precipitation for 2060

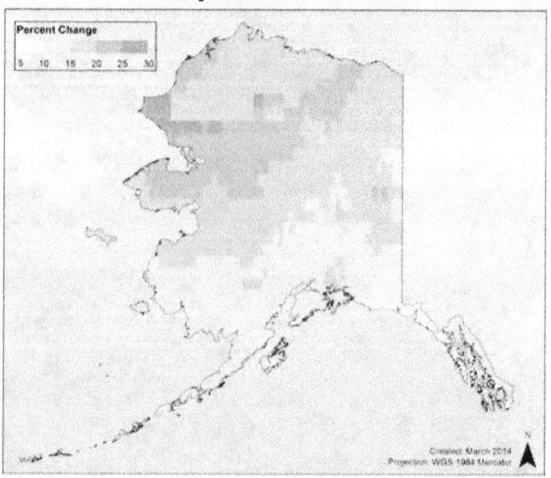

Percent Change
5   10   15   20   25   30

United States Environmental Protection Agency

CLIMATE READY
WATER UTILITIES

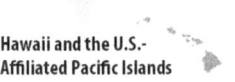

Hawaii and the U.S.-Affiliated Pacific Islands

Climate Region Brief > HAWAII and the U.S.-AFFILIATED PACIFIC ISLANDS

Return to Introduction

Climate change for the U.S.-afilliated islands in the Pacific Ocean is projected to continue to follow already observable trends. Temperature rise, shifts in precipitation patterns and timing and altered hydrologic cycles can be expected due to climate change. The following statements, drawn from U.S. Global Change Research Program assessments (USGCRP 2009, USGCRP 2014), are based on projections for climate conditions at the end of the 21st century – using both high and low emissions scenarios (IPCC 2000).

OBSERVED AND PROJECTED CHANGES

- Ocean surface temperature has increased by as much as 3.6°F since the 1950s. Projections for the rest of this century suggest increases in air and ocean surface temperatures in the Pacific Ocean.

- Average annual precipitation, average stream discharge and stream base flow have been trending downward for nearly a century, especially in recent decades, but with high variability.

- Rainfall during summer months, traditionally the drier part of the year, is expected to increase 5% by the end of the century in the Pacific and may result in unusual summer flooding.

- In Hawaii, decreases in precipitation are projected for the northern islands under the lower emissions scenario through the end of the 21st century and under the higher emissions scenario through the middle of the 21st century. Increases in precipitation are projected for the southern islands under the higher emissions scenario.

- Changes in weather patterns are projected to cause an increase in the frequency and intensity of extreme storm events, sea-level rise, coastal erosion, coral reef bleaching, ocean acidification and contamination of freshwater resources by saltwater.

- Islands and other low-lying coastal areas will be at increased risk from coastal inundation due to sea-level rise and storm surge.

- Hurricane (typhoon) wind speeds and rainfall rates are likely to increase, which, combined with sea-level rise, is expected to cause higher storm surge levels.

GROUP		DW	WW
Drought	Reduced groundwater recharge	⬤	
	Lower lake & reservoir levels	⬤	
	Changes in seasonal runoff & loss of snowpack	⬤	
Water Quality Degradation	Low flow conditions & altered water quality		⬤
	Saltwater intrusion into aquifers	⬤⬤	
	Altered surface water quality	⬤	⬤
Floods	High flow events & flooding	⬤⬤	⬤⬤
	Flooding from coastal storm surges	⬤⬤	⬤⬤
Ecosystem Changes	Loss of coastal landforms / wetlands	⬤⬤	⬤⬤
	Increased fire risk & altered vegetation	⬤	⬤
Service Demand & Use	Volume & temperature challenges	⬤	⬤
	Changes in agricultural water demand	⬤	
	Changes in energy sector needs	⬤	
	Changes in energy needs of utilities	⬤⬤	⬤⬤

Click on a group name above to read more about these impacts or click on a water drop above to read more about a specific impact.

⬤⬤ = Particularly relevant to Hawaii and the U.S. Affiliated Pacific Islands
⬤ = Somewhat relevant

Projections for the rest of this century suggest increases in air and ocean surface temperatures in the Pacific Ocean. These figures to the right display projected increases in the annual mean temperature of Hawaiian Islands for three future time periods under both high and low emissions scenarios. Simulations indicate a statistically significant increase in annual mean temperature for all future time periods (Kunkel et al. 2013).

Simulated Change in Annual Mean Temperature

Preliminary CMIP5 projections show a clear tendency for increased wet season rainfall in the Western Pacific, with an overall trend towards fewer, extremely high rainfall events. The figure to the right shows a tendency for increased wet season rainfall in the Western Pacific islands through the end of the century as the climate warms (Keener et al. 2013).

Mean Wet Season (June-September) Rainfall for the Tropical Western Pacific
(GFDL CM3 Model, RCP8.5 Simulator)

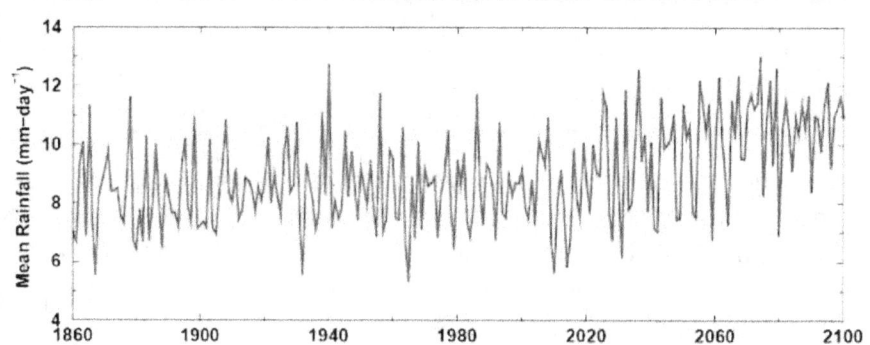

United States
Environmental Protection
Agency

CLIMATE READY
WATER UTILITIES
⌁EPA

Coasts

Climate Region Brief > COASTS

Return to Introduction

Climate change for U.S. coastal areas is projected to continue to follow already observable trends. Temperature rise, shifts in precipitation patterns and timing and altered hydrologic cycles can be expected due to climate change. The following statements, drawn from U.S. Global Change Research Program assessments (USGCRP 2009, USGCRP 2014), are based on projections for climate conditions at the end of the 21st century – using both high and low emissions scenarios (IPCC 2000).

OBSERVED AND PROJECTED CHANGES

- Coastal waters are very likely to continue to warm by as much 4 to 8 °F in this century, both in summer and winter.

- More spring runoff and warmer coastal waters will increase the seasonal reduction in oxygen and increase the area and intensity of coastal dead zones in places such as the northern Gulf of Mexico and the Chesapeake Bay.

- Significant sea-level rise and storm surge will erode shorelines and adversely affect coastal cities and ecosystems. Sea level has risen along most of the coast over the last 50 years, and will rise more in the future. Sea level is projected to rise another 1 to 4 feet by 2100.

- Sea-level rise is expected to increase saltwater intrusion into coastal freshwater aquifers, making some unusable without desalination.

GROUP		DW	WW
Drought	Reduced groundwater recharge	◌	
	Lower lake & reservoir levels	◌	
	Changes in seasonal runoff & loss of snowpack	◌◌	
Water Quality Degradation	Low flow conditions & altered water quality		◌◌
	Saltwater intrusion into aquifers	◌	
	Altered surface water quality	◌	◌
Floods	High flow events & flooding	◌◌	◌◌
	Flooding from coastal storm surges	◌◌	◌◌
Ecosystem Changes	Loss of coastal landforms / wetlands	◌◌	◌◌
	Increased fire risk & altered vegetation	◌	◌
Service Demand & Use	Volume & temperature challenges	◌◌	◌
	Changes in agricultural water demand	◌	
	Changes in energy sector needs	◌	
	Changes in energy needs of utilities	◌◌	◌◌

Click on a group name above to read more about these impacts or click on a water drop above to read more about a specific impact.

◌◌ = Particularly relevant to coasts   ◌ = Somewhat relevant

- The intensity, frequency and duration of North Atlantic hurricanes has increased in recent decades, and the intensity of these storms is likely to increase during this century.

Continued on page 2

Over the past 100 years, sea level has increased 1.2 feet in New York City. This graph shows that the observed rise in sea level at the Battery in New York City has significantly exceeded the global average of 8 inches over the past century, increasing the risk of impacts to critical urban infrastructure in low-lying areas (NPCC 2010).

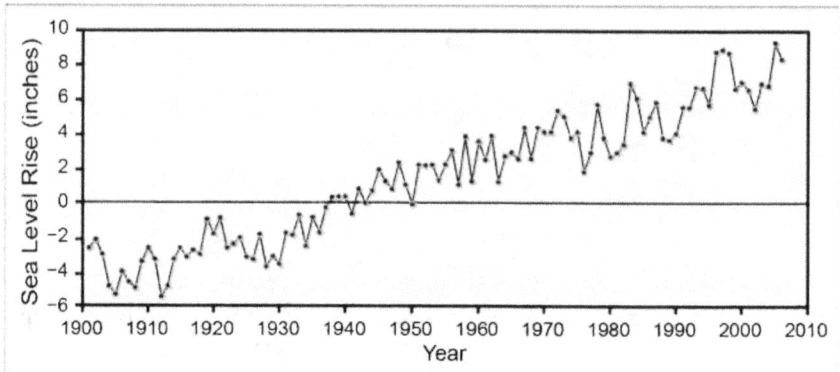

Observed Sea-Level Rise in New York City

Vulnerability to Sea Level Rise

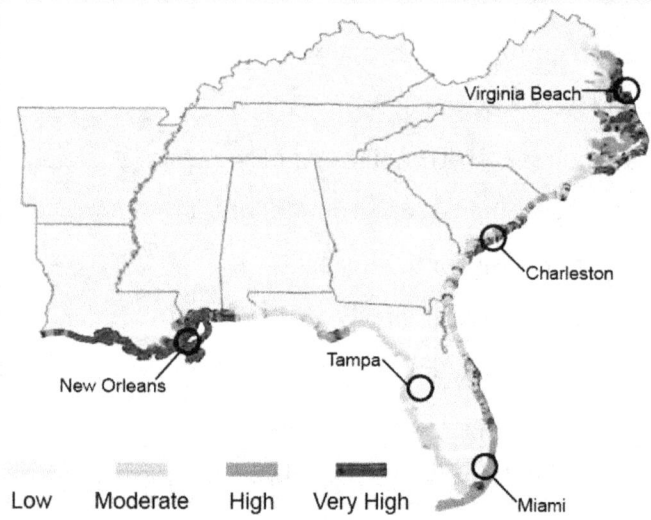

Significant sea-level rise and storm surge will erode shorelines and adversely affect coastal cities and ecosystems in the Southeast. The figure to the left shows the relative risk that physical changes will occur as sea level rises. The Coastal Vulnerability Index used here is calculated based on tidal range, wave height, coastal slope, shoreline change, landform and processes, and historical rate of relative sea level rise. The approach combines an assessment of a coastal system's susceptibility to change with its natural ability to adapt to changing environmental conditions, and yields a relative measure of the system's natural vulnerability to the effects of sea-level rise (Hammar-Klose and Thieler 2001, USGCRP 2014).

 **United States Environmental Protection Agency**

CLIMATE READY
WATER UTILITIES

Group: DROUGHT (DW)

Return to Introduction

Observed data indicate that drought intensity and frequency have been increasing in the United States during the last few decades, especially in much of the West. Average values of the Palmer Drought Severity Index from 2000-2010 indicated the most severe average drought of any decade on record. Summer droughts are expected to intensify in most regions of the United States (USGCRP 2014). The impacts to water utilities from drought associated with climate change may be driven or forced by changing water levels in aquifers and reservoirs, loss of snowpack and reductions in surface water flows. Clicking on the drinking water icon next to each impact name will bring you to that particular Strategy Brief. Clicking on the Green Infrastructure or Water Demand Management icon will bring you to that Sustainability Brief.

Reduced Groundwater Recharge

Reduced precipitation and higher loss of water from plants and evaporation due to higher temperatures will decrease surface water supplies and groundwater recharge, especially impacting utilities that rely on groundwater supplies. Review this brief to learn more about how the Inland Empire Utilities Agency (IEUA) used stormwater capture and water recycling to counteract the effects of reduced groundwater recharge and how Tucson Water has constructed a large-scale recharge and recovery system to secure its water supply through 2050.

Lower Lake and Reservoir Levels

Decreases in mean annual precipitation and higher loss of water from vegetation and evaporation due to higher temperatures will lead to lower levels in the lakes and reservoirs that water utilities rely on for surface water supplies. These lower levels may make it difficult to meet water demands, especially in the summer months, and may drop water levels below intake infrastructure. Review this brief to learn more about how Southern Nevada Water Authority (SNWA) uses aggressive conservation practices and new construction to address falling water levels in Lake Mead.

Changes in Runoff and Loss of Snowpack

Increased temperatures and shifting precipitation patterns will alter seasonal runoff and storage of water in snowpack. These changes in water supply could strain the capacity of reservoirs to hold larger and earlier peak runoff flows, cause shortages in the summer due to longer duration of the warmer and drier season and compromise biodiversity goals (e.g., managing cold-water fish, such as salmon and trout). Lower annual precipitation will lead to lower streamflow in many locations, which may lead to diminished water quality. Diminished water quality in receiving waters may lead to more stringent requirements for wastewater discharges, leading to higher treatment costs and the need for capital improvements. Review this brief to learn more about how the Portland Water Bureau is considering expanding its groundwater supply or surface water storage to offset the impacts seasonal runoff changes will have on water supply and how East Bay Municipal Utility District (EBMUD) used results of a "bottom up" sensitivity analysis to plan for impacts related to projected earlier runoff.

Click to left of name to check off options for consideration; $'s ($-$$$) indicate relative costs
Click name of any option to review more information in the Glossary
No Regrets options - actions that would provide benefits to the utility under current climate conditions as well as any future changes in climate. For more information on No Regrets options, see Page 11 in the Introduction.
Click on the 💡, 🌱 or 💧 icon to review the relevant Sustainability Brief.

ADAPTATION OPTIONS

✓	PLANNING	COST
⊛	Develop models to understand potential water quality changes (e.g., increased turbidity) and costs of resultant changes in treatment.	$
⊛	Incorporate monitoring of groundwater conditions and climate change projections into groundwater models.	$

Continued on page 2

✓ **PLANNING** (continued)	**COST**
⊛ Use hydrologic models to project runoff and incorporate model results during water supply planning.	$
⊛ Conduct climate change impacts and adaptation training for personnel.	$
✋ Participate in community planning and regional collaborations related to climate change adaptation.	$-$$

✓ **OPERATIONAL STRATEGIES**	**COST**
⊛ Monitor current weather conditions, including precipitation and temperature.	$
⊛ Monitor surface water conditions, including river discharge and snowmelt.	$
⬤ Finance and facilitate systems to recycle water, including use of greywater in homes and businesses.	$$-$$$
Practice conjunctive use (i.e., optimal use of surface water and groundwater).	$$-$$$
⬤ Reduce agricultural and irrigation water demand by working with irrigators to install advanced equipment (e.g., drip or other micro-irrigation systems with weather-linked controls).	$$-$$$
⬤ Practice demand management through communication to public on water conservation actions.	$
💡⬤ Practice water conservation and demand management through water metering, leak detection and water loss monitoring, rebates for water conserving appliances/toilets and/or rainwater harvesting tanks.	$-$$

✓ **CAPITAL/INFRASTRUCTURE STRATEGIES**	**COST**
✋ Acquire and manage ecosystems, such as forested watersheds, vegetation strips and wetlands, to regulate runoff.	$$$
Build infrastructure needed for aquifer storage and recovery, either for seasonal storage or longer-term water banking, (e.g., recharge canals, recovery wells).	$$$
⊛⬤ Diversify options to complement current water supply, including recycled water, desalination, conjunctive use and stormwater capture.	$$$
⊛ Expand current resources by developing regional water connections to allow for water trading in times of service disruption or shortage.	$$-$$$
⊛ Increase water storage capacity, including silt removal to expand capacity at existing reservoirs and construction of new reservoirs and/or dams.	$$-$$$
⊛ Increase or modify treatment capabilities to address treatment needs of marginal water quality in new sources.	$$$
Retrofit intakes to accommodate lower water levels in reservoirs and decreased late season flows.	$$-$$$
Build or expand infrastructure to support conjunctive use.	$$$
💡⬤ Build systems to recycle wastewater for energy, industrial, agricultural or household use.	$$$

 United States
Environmental Protection
Agency

CLIMATE READY
WATER UTILITIES
⚘EPA

REDUCED GROUNDWATER RECHARGE (DW)

Return to Introduction

Reduced precipitation, decline in runoff, projected loss of snowpack, increased loss of water from vegetation and evaporation due to higher temperatures will not only lead to decreases in surface water supplies; these changes will also lead to decreased groundwater recharge, impacting utilities that rely on groundwater supplies. Decreases in available surface water have already resulted in higher groundwater use in some areas, including the Central Valley of California, where during the 2006-2009 drought, groundwater storage declined by an estimated 24 km^3 to 31 km^3, equivalent to the storage capacity in Lake Mead (Famiglietti et al. 2011).

CLIMATE INFORMATION

- Between 1895 and 2011, mean annual precipitation in the United States increased by close to 2 inches, but there have been important regional and seasonal differences. Over the past few decades, decreases in annual precipitation have been observed in Hawaii and in parts of the Southeast and Southwest (USGCRP 2014).

- Similarly, climate projections indicate differences in precipitation trends by region. For the Southwest, climate models project continued decreases in mean annual precipitation in this century (Orlowsky and Seneviratne 2012). Seasonal precipitation projections for winter and spring in the Southwest also show a decreasing trend. However, in the northern part of the United States, winter and spring precipitation are projected to increase (USGCRP 2014).

- Many southwestern and western watersheds are experiencing increasingly drier conditions. Even larger runoff reductions (10 to 20%) are expected over some of these watersheds in the next 50 years (Cayan et al. 2010).

Click to left of name to check off options for consideration; $'s (**$-$$$**) indicate relative costs
Click name of any option to review more information in the Glossary
⭐ **No Regrets options** - actions that would provide benefits to the utility under current climate conditions as well as any future changes in climate. For more information on No Regrets options, see Page 11 in the Introduction.
Click on the 💡 🌊 or 🖐 icon to review the relevant Sustainability Brief.

ADAPTATION OPTIONS

✓ PLANNING	COST
⭐ Incorporate monitoring of groundwater conditions and climate change projections into groundwater models.	$
⭐ Conduct climate change impacts and adaptation training for personnel.	$
💡 Participate in community planning and regional collaborations related to climate change adaptation.	$-$$

✓ OPERATIONAL STRATEGIES	COST
⭐ Monitor current weather conditions, including precipitation and temperature.	$
🖐 Finance and facilitate systems to recycle water, including use of greywater in homes and businesses.	$$ $$$
Practice conjunctive use (i.e., optimal use of surface water and groundwater).	$$-$$$
🖐 Reduce agricultural and irrigation water demand by working with irrigators to install advanced equipment (e.g., drip or other micro-irrigation systems with weather-linked controls).	$$-$$$
🖐 Practice demand management through communication to public on water conservation actions.	$
💡🖐 Practice water conservation and demand management through water metering, leak detection and water loss monitoring, rebates for water conserving appliances/toilets and/or rainwater harvesting tanks.	$-$$

 Continued on page 2

✓	CAPITAL/INFRASTRUCTURE STRATEGIES	COST
	Build infrastructure needed for aquifer storage and recovery, (either for seasonal storage or longer-term water banking), (e.g., recharge canals, recovery wells). *(See example 2 below)*	$$$
⭐	Expand current resources by developing regional water connections to allow for water trading in times of service disruption or shortage.	$$-$$$
⭐ ⬤	Diversify options to complement current water supply, including recycled water, desalination, conjunctive use and stormwater capture. *(See examples 1 and 2 below)*	$$$
⭐	Increase water storage capacity, including silt removal to expand capacity at existing reservoirs and construction of new reservoirs and/or dams.	$$-$$$
	Build or expand infrastructure to support conjunctive use.	$$$
💡 ⬤	Build systems to recycle wastewater for energy, industrial, agricultural or household use.	$$$

EXAMPLE 1

The Inland Empire Utilities Agency (IEUA) is a wholesale water and wastewater service provider in Southern California's Riverside County. Currently, the IEUA region receives more than half of its average water needs from groundwater sources (primarily the underlying Chino Basin Aquifer), about a quarter from Northern California via a large intrastate water distribution system (the California State Water Project) and the rest from surface water and a rapidly expanding recycled water system. An analysis, based on projections from an ensemble of 21 climate models, showed that winter precipitation between 2000 and 2030 could change from -27% to +19%. However, due to the potentially hotter and drier conditions, outdoor water demand could increase by 11% by 2040 assuming constant land use patterns, demand factors and water supply variability. There may be decreasing sustainable groundwater yields of up to -15% by 2040 (Groves et al. 2008). IEUA conducted a robust decision-making analysis with the goal of adopting adaptation measures that would not exceed $3.75 billion. In response to the results of this analysis, the utility decided to accelerate expansion of its dry-year-yield program (i.e., groundwater recharge using stormwater) as well as implementation of its water recycling efforts. These efforts involve the reuse of tertiary treated wastewater for groundwater recharge and other functions. In total, the water recycling plan calls for an increase in recycled water from 9.9 million m³/year in 2005 to 85 million m³/year in 2025 (Groves et al. 2008, Lembert and Groves 2010). Implementing these measures will help counteract the effects of reduced groundwater recharge due to climate change.

EXAMPLE 2

Before the year 2001, Tucson, Arizona was the largest city in the country completely dependent on groundwater, a stressed and potentially shrinking resource, for its water supply. To ensure a more reliable and sustainable supply, Tucson Water developed a framework in the 1990s that outlines strategies to reduce dependence on groundwater. Tucson Water is allocated 144,191 acre-feet of water per year from the Colorado River through the construction of the Central Arizona Project (CAP), a canal system that brings water from the Colorado River into Arizona. As a part of this framework, Tucson Water began constructing a large scale recharge/recovery system for the water purchased from its CAP allocation in the late 1990s. The system uses allocated water to recharge underground wells from which the supply is then pumped and distributed to customers. Tucson Water can currently recharge its entire CAP allocation, which exceeds its annual demand – allowing the utility to store almost 45,000 acre-feet of water per year for future use.

Recharging groundwater with allocated water from the Colorado River provides Tucson with complete system redundancy with no reliance on stressed groundwater aquifers. Tucson Water projects that it can maintain reliable service to customers through 2050 using only the recharged water, even considering a reduced allocation of water from the Colorado River (Tucson 2012). The stored groundwater is also useful in meeting peak demand during hot summer months which may continue to grow as summer temperatures rise. In the future, Tucson Water plans to construct additional spreading basins and production wells to provide storage and wheeling opportunities to local and regional CAP partners. Tucson Water will also begin implementation of its Recycled Water Program to make underutilized reclaimed water available to meet future potable needs (Tucson 2013).

 United States
Environmental Protection
Agency

CLIMATE READY
WATER UTILITIES
&EPA

LOWER LAKE AND RESERVOIR LEVELS (DW)

Return to Introduction

Reductions in lake and reservoir levels may occur due to the combined impacts of decreased mean annual precipitation, reductions in runoff and higher loss of water from vegetation and evaporation due to higher temperatures. These surface water resources are critical for water utilities that lack other viable sources. Recent data indicate that lake levels for three of the Great Lakes (Superior, Michigan and Ontario) have been below their long-term averages for much of 2000-2010 (NOAA 2012). Summer drought has left both Lake Michigan and Lake Ontario water levels approximately 1 foot below the long-term average. These lower levels may exacerbate the ability of utilities to meet water demands, especially during summer months, and in some cases may drop water levels below intake infrastructure.

CLIMATE INFORMATION

- Between 1895 and 2011, mean annual precipitation increased by more than 2 inches in the United States, but important regional and seasonal differences have been observed. Over the past few decades, decreases in annual precipitation have been seen in Hawaii as well as parts of the Southeast and Southwest.

- Climate models project that decreases in annual precipitation in the Southwest will continue in this century (Orlowsky and Seneviratne 2012). In general, increased winter and spring precipitation is projected for the northern part of the U.S., while decreased winter and spring precipitation is projected for the Southwest (USGCRP 2014). For utilities in these areas, declining precipitation may translate directly into decreased volumes in source lakes and reservoirs.

- Many southwestern and western watersheds are currently experiencing increasingly drier conditions. Even larger reductions in runoff (10% to 20%) are expected over some of these watersheds in the next 50 years. Runoff reductions can impact surface water supplies and result in decreased water volumes (Cayan et al. 2012).

Click to left of name to check off options for consideration; $'s (**$-$$$**) indicate relative costs
Click name of any option to review more information in the Glossary
No Regrets options - actions that would provide benefits to the utility under current climate conditions as well as any future changes in climate. For more information on No Regrets options, see Page 11 in the Introduction.
Click on the ⊙, ⊙ or ⊙ icon to review the relevant Sustainability Brief.

ADAPTATION OPTIONS

✓ PLANNING	COST
⊙ Conduct training for personnel in climate change impacts and adaptation.	$
⊙ Participate in community planning and regional collaborations related to climate change adaptation. *(See example below)*	$-$$

✓ OPERATIONAL STRATEGIES	COST
⊙ Monitor current weather conditions, including precipitation and temperature.	$
⊙ Finance and facilitate systems to recycle water, including use of greywater in homes and businesses.	$$-$$$
Practice conjunctive use (i.e., optimal use of surface water and groundwater).	$$-$$$
⊙ Reduce agricultural and irrigation water demand by working with irrigators to install advanced equipment (e.g., drip or other micro-irrigation systems with weather-linked controls). *(See example below)*	$$-$$$
⊙ Practice demand management through communication to public on water conservation actions. *(See example below)*	$

Continued on page 2

 United States
Environmental Protection
Agency

 CLIMATE READY
WATER UTILITIES
♻EPA

Adaptation Strategies Guide for Water Utilities
ABOUT THIS GUIDE

Climate change presents several **challenges** to drinking water and wastewater utilities, including increased frequency and duration of droughts, floods associated with intense precipitation events and coastal storms, degraded water quality, wildfires and coastal erosion and subsequent changes in demand for services. While these impacts have been documented in numerous publications, finding the right information for your **type of utility** or **geographic region** can be difficult and sometimes overwhelming. Therefore, the goals of the Adaptation Strategies Guide are (1) to provide drinking water and wastewater utilities with a basic understanding of how climate change can impact utility operations and missions, and (2) to provide examples of different actions utilities can take (i.e., adaptation options) to prepare for these impacts.

The climate information included in this Guide (identified as **projected and observed change** statements throughout the document) has been updated to reflect the findings from the U.S. Global Change Research Program (USGCRP) 2014 Report. For more information on the changes to the climate data found in the USGCRP 2014 Report and this Guide, see the "Updated Climate Information" section below. Global climate research, conducted by international research groups, has generated projections of future climate conditions based on historical climate data (i.e., temperature, precipitation and sea level) as well as simulations based on scientific understanding of atmospheric processes. These groups and other research institutions have translated and "downscaled" projections from global models to produce projections at national, regional and local scales. In many cases, projected changes in climate may generate specific impacts or challenges for drinking water and wastewater utilities that are described within this Guide. This process of translating global climate projections into the challenges that drinking water and wastewater utilities may face is outlined in **Figure 1.1** and is described in greater detail in the 2014 USGCRP report and in other reports being released as part of the National Climate Assessment process (http://www.globalchange.gov/what-we-do/assessment).

Adapting your utility system and operations to climate change challenges requires consideration and planning. However, adaptation planning is not necessarily a new effort distinct from other utility practices. Since adaptation strategies often provide multiple benefits, adaptation planning can be integrated into existing efforts for emergency response planning, capacity development, capital investment planning, water supply and demand planning, conservation practices, sustainability goals and infrastructure maintenance.

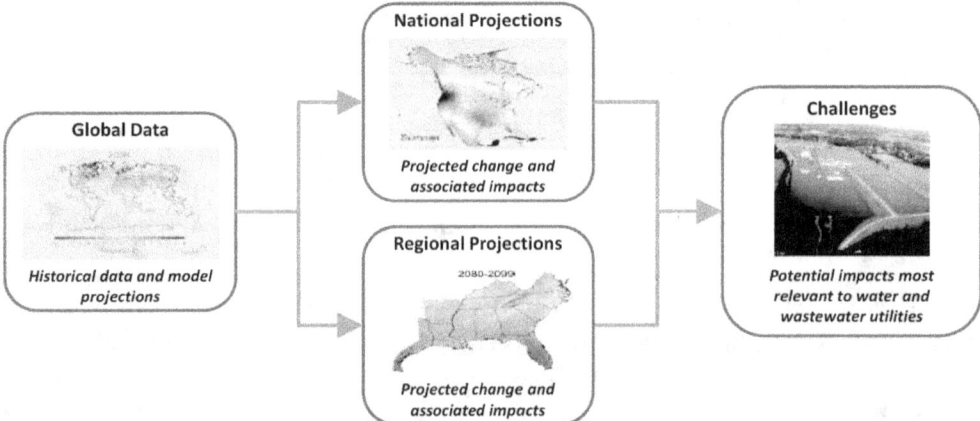

Figure 1.1. Scheme to translate climate data and model outputs into challenges for water utilities. Model run images from USGCRP (2009).

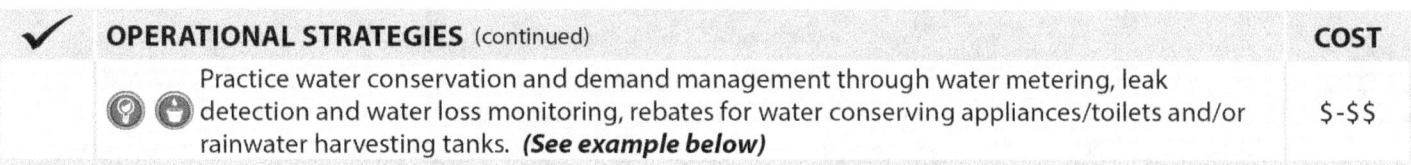

OPERATIONAL STRATEGIES (continued)	**COST**
Practice water conservation and demand management through water metering, leak detection and water loss monitoring, rebates for water conserving appliances/toilets and/or rainwater harvesting tanks. **(See example below)**	$-$$

CAPITAL/INFRASTRUCTURE STRATEGIES	**COST**
Acquire and manage ecosystems, such as forested watersheds, vegetation strips and wetlands, to regulate runoff.	$$$
Build infrastructure needed for aquifer storage and recovery, (either for seasonal storage or longer-term water banking), (e.g., recharge canals, recovery wells).	$$$
Diversify options to complement current water supply, including recycled water, desalination, conjunctive use and stormwater capture.	$$$
Expand current resources by developing regional water connections to allow for water trading in times of service disruption or shortage.	$$-$$$
Increase water storage capacity, including silt removal to expand capacity at existing reservoirs and construction of new reservoirs and/or dams.	$$-$$$
Increase or modify treatment capabilities to address treatment needs of marginal water quality in new sources.	$$$
Retrofit intakes to accommodate lower water levels in reservoirs. **(See example below)**	$$-$$$
Build or expand infrastructure to support conjunctive use.	$$$
Build systems to recycle wastewater for energy, industrial, agricultural or household use.	$$$

EXAMPLE

The Southern Nevada Water Authority (SNWA) and its member agencies supply water to approximately 2 million people and more than 40 million annual visitors. The SNWA currently draws about 90 percent of the community's water supply from the Colorado River via Lake Mead, the largest man-made reservoir in the United States. During the past decade, the impact of drought has caused Lake Mead elevations to decline by more than 100 feet, representing a storage loss of more than 4 trillion gallons. The SNWA is concerned about further reductions in streamflow from climate change and increased demands. A study conducted by the Bureau of Reclamation and the Colorado River Basin States projected a 3.2 million acre-foot annual imbalance between supply and demand for the Colorado River Basin by 2060 (Bureau of Reclamation 2012). For the SNWA, reduced Colorado River streamflow would result in lower levels in Lake Mead, the potential loss of the ability to withdraw water from existing intakes, reduced water quality at withdrawal locations, and increased power requirements to pump water a greater vertical distance.

To ensure a reliable supply for residents and visitors into the future, the SNWA launched an extensive water conservation program more than a decade ago; It is also investing in significant infrastructure enhancements related to its Lake Mead intakes. Demand management practices (i.e., education, incentives, regulation and rates) have reduced consumptive water use by 32% since 2000, even as the population has increased by nearly half a million. Examples of successful strategies include: incentives for homeowners and commercial properties that convert turf to water efficient landscapes; working with landscapers in the area to provide them with water-efficient irrigation technology; rebates on pool covers; and time/day restrictions on landscape irrigation, including for commercial customers. Infrastructure enhancements include the completion of a second intake, ongoing construction of a third deeper intake in Lake Mead, additional water treatment capacity using ozone, and distribution system expansions.

The SNWA is also using EPA's Climate Resilience Evaluation and Awareness Tool (CREAT) to evaluate a number of physical adaptation measures to address the impacts related to declining lake levels. These options include infrastructure improvements to ensure operability at lower lake levels and constructing a new intake that can withdraw water from deeper in the lake where the water is cooler and of higher quality.

EXAMPLE (continued)

Despite the agency's successful conservation strategies, reservoir levels may continue to decline. Even if SNWA stopped withdrawing water entirely for a year, Lake Mead would only rise by approximately 3 feet, given its current elevation, which would do little to offset the 100 feet of decline that has been seen in the past 14 years. Recognizing this, SNWA has made it a priority to work with the other states that rely on the Colorado River and Mexico to develop innovative solutions through key partnerships (Bureau of Reclamation 2012).

For more information see: http://www.snwa.com/ws/resource_plan.html

 United States
Environmental Protection
Agency

CLIMATE READY
WATER UTILITIES
♻EPA

CHANGES IN SEASONAL RUNOFF & LOSS OF SNOWPACK (DW) Return to Introduction

Increased temperatures and shifting precipitation patterns will alter seasonal runoff and storage of water in snowpacks. These disruptions in water supply could strain the capacity of reservoirs to hold larger and earlier peak runoff flows, cause shortages in the summer due to the longer hot and dry season and compromise biodiversity goals (e.g., managing cold-water fish, such as salmon and trout).

CLIMATE INFORMATION

- Over the last 50 years, there have been widespread temperature-related reductions in snowpack in the West, particularly at lower elevations. In both the West and the Northeast, there has been a transition to more rain and less snow.

- Declines in spring snowpack, earlier snowmelt-fed streamflow and more precipitation falling as rain instead of snow in much of the western United States have been observed since 1950. In some locations in the West, half of the annual flow has arrived 5 to 20 days earlier each year from 2001-2010, compared to the average from the second half of the 20th century (USGCRP 2014). For the Northeast, runoff from snowmelt is projected to occur earlier in the year. By the end of the century, in some cases spring runoff in the West may occur up to 60 days earlier, while in the Northeast, it could advance 14 days (USGCRP 2009).

- Between 2001 and 2010, streamflow totals in many river basins in the Southwest were 5% to 37% lower than 20th century average flows (Hoerling et al. 2012). Annual runoff and streamflow are projected to decline in the Southwest and Southeast. Mean runoff declines are projected throughout the year and especially from November to May for some river basins in the Southeast (USGCRP 2014).

- Many southwestern and western watersheds are experiencing increasingly drier conditions with projected runoff reductions ranging from 10 to 20% over some watersheds in the next 50 years (Cayan et al. 2010).

- In California's Sierra Nevada Mountains, snowpack reductions are projected to range from 25% to 40% by 2050, leading to a loss of snowpack storage from an average of 15 million acre-feet to an estimated 9 to 10.5 million acre-feet per year. By 2090, assuming an increase in mean temperatures of 3.8 °F, the watershed upstream of the San Francisco estuary could lose 50% of its April snowpack (Standish-Lee and Lecina 2008).

ADAPTATION OPTIONS

Click to left of name to check off options for consideration; $'s (**$-$$$**) indicate relative costs

Click name of any option to review more information in the Glossary

⊛ **No Regrets options** - actions that would provide benefits to the utility under current climate conditions as well as any future changes in climate. For more information on No Regrets options, see Page 9 in the Introduction.

Click on the 💡, 💧 or ☕ icon to review the relevant Sustainability Brief.

✓ PLANNING	COST
⊛ Develop models to understand potential water quality changes (e.g., increased turbidity) and costs of resultant changes in treatment.	$
⊛ Use hydrologic models to project runoff and incorporate model results during water supply planning. **(See examples 1 and 2 below)**	$
⊛ Conduct training for personnel in climate change impacts and adaptation strategies.	$
⊛ Participate in community planning and regional collaborations related to climate change adaptation.	$-$$

✓ OPERATIONAL STRATEGIES	COST
⊛ Monitor current weather conditions, including precipitation and temperature.	$

Continued on page 2

✓	**OPERATIONAL STRATEGIES**	**COST**
⊛ Monitor surface water conditions, including river discharge and snowmelt.	$
🖐 Finance and facilitate systems to recycle water, including use of greywater in homes and businesses.	$$-$$$
Practice conjunctive use (i.e., optimal use of surface water and groundwater). *(See examples 1 and 2 below)*	$$-$$$
⊙ Reduce agricultural and irrigation water demand by working with irrigators to install advanced equipment (e.g., drip or other micro-irrigation systems with weather-linked controls).	$$-$$$
💡🖐 Practice water conservation and demand management through water metering, leak detection and water loss monitoring, rebates for water conserving appliances/toilets and/or rainwater harvesting tanks.	$-$$

✓	**CAPITAL/INFRASTRUCTURE STRATEGIES**	**COST**
⊕ Acquire and manage ecosystems, such as forested watersheds, vegetation strips, and wetlands, to regulate runoff.	$$$
Build infrastructure needed for aquifer storage and recovery, (either for seasonal storage or longer-term water banking), (e.g., recharge canals, recovery wells).	$$$
⊛⊙ Diversify options to complement current water supply, including recycled water, desalination, conjunctive use, and stormwater capture. *(See example 2 below)*	$$$
⊛ Expand current resources by developing regional water connections to allow for water trading in times of service disruption or shortage.	$$-$$$
⊛ Increase water storage capacity, including silt removal to expand capacity at existing reservoirs and construction of new reservoirs and/or dams. *(See example 1 below)*	$$-$$$
Retrofit intakes to accommodate decreased flow in source waters.	$$-$$$
Build or expand infrastructure to support conjunctive use.	$$$

EXAMPLE 1

The Portland Water Bureau supplies water to approximately 800,000 people in the Portland, Oregon metropolitan area, delivering 40 billion gallons per year. The primary water source is the Bull Run Watershed, and there are two reservoirs with a combined total storage capacity of 10 billion gallons. Precipitation over the watershed ranges from 59 inches to more than 80 inches per year – most falling during the winter months. The greatest challenge for the utility is supplying water during the summer months, when demand (220 million gallons per day) is double the average daily use. The utility's secondary water source is groundwater located along the south shore of the Columbia River.

The utility generated future scenarios of water supply and demand using four different climate model projections and regional population growth projections. Results such as increased winter precipitation, earlier snowmelt and drier summers were consistent across the models. The main concern is not a reduction in annual precipitation but seasonal changes in runoff: spring runoff may increase by 15%, followed by late spring reductions in runoff of 30%. The analysis suggests that reduced summer precipitation combined with increased seasonal demand may lead to decreased reliability of supply, unless additional infrastructure is provided. The impact would result in a 2.8–5.4 billon gallon decrease in reservoir storage. To ameliorate this, the utility is considering expanding groundwater supply or surface water storage. The latter would allow for the sustainability of the emergency groundwater supply. Besides storage augmentation, other measures that are being considered include conjunctive use strategies that coordinate the optimal use of existing surface and groundwater supplies, including use of aquifer storage and recovery (Miller and Yates 2005).

EXAMPLE 2

The East Bay Municipal Utility District (EBMUD) supplies water to 1.3 million people and provides wastewater service to 650,000 people in portions of Alameda and Contra Counties in the San Francisco Bay Area. The primary water source is the Mokelumne River Watershed on the western slopes of the Sierra Nevada Mountains. EBMUD's average water demand is 210 million gallons per day (77 billion gallons per year) and its seven raw water reservoirs have a storage capacity of 766,740 acre-feet (250 billion gallons). Annually, precipitation in the Mokelumne River Watershed averages approximately 48 inches. EBMUD must manage its water supply to meet multiple objectives, including: municipal supply, streamflow regulation, fishery requirements, flood control, downstream water obligations, recreation and hydropower generation. EBMUD also has a dry year supplemental supply from the Sacramento River.

The utility performed a "bottom up" sensitivity analysis to evaluate the impact of a 4°C increase in temperature and 20% decrease in precipitation on its carryover storage, flood control releases, customer rationing and river temperature. The analysis found that earlier runoff as a result of increasing temperature could increase flood control releases in up to 60% of the years simulated, and carryover storage could decrease in up to 56% of the years simulated. Additionally, carryover storage decreased 3% to 12% in the years simulated. The increase in river temperature varied between 0.3°C to 3.5°C and depending on whether it was a dry or wet year. These results were included in EBMUD's Water Supply Master Plan that looks out to the year 2040. The plan includes alternative strategies for increasing water conservation and water recycling, utilizing its supplemental supply from the Sacramento River in dry years, and investigating groundwater and regional water projects.

 United States
Environmental Protection
Agency

CLIMATE READY
WATER UTILITIES
♨EPA

Group: WATER QUALITY DEGRADATION (DW/WW)

Return to Introduction

Changes in water quality associated with climate change may be driven or forced by saline intrusion into aquifers and altered surface water quality. Clicking on either the drinking water or wastewater icon next to each impact name will bring you to that particular Strategy Brief. Clicking on the Energy Management or Green Infrastructure icon will bring you to that Sustainability Brief.

Low Flow Conditions and Altered Water Quality

Many areas are projected to receive less annual total precipitation concentrated in fewer, more extreme rainfall events. Lower annual precipitation will lead to lower streamflows in many locations, which may lead to diminished water quality. Turbidity from sediment washing downstream following storm events also impacts water quality, particularly in areas where fires have diminished the ability of landscapes to hold soil. Diminished water quality in receiving waters may lead to more stringent requirements for wastewater discharges and impacts to ecosystems that are sensitive to temperature. Review this brief to learn more about how Spartanburg Water coordinates reservoir releases with the wastewater system to limit water quality issues associated with wastewater discharge into water bodies.

Saltwater Intrusion into Aquifers

Projected sea-level rise, combined with higher water demand from coastal communities, can lead to saltwater intrusion in both coastal groundwater aquifers and estuaries. This combination may reduce water quality and increase treatment costs for water treatment facilities drawing from coastal aquifers or surface water intakes in tidal estuaries near the saltwater line. Desalination plants may have to treat water with higher salt content, which would also increase costs. Review this brief to learn more about how the Los Angeles County Flood Control District constructed groundwater injection barriers to block saltwater intrusion.

Altered Surface Water Quality

Climate models project that the average annual temperature in the United States, as well as the number of extreme hot days, will increase. Higher temperatures can lead to algal blooms, which compromise source water quality and may require more advanced treatment. These water quality impacts will drive the need for additional treatment processes for drinking water utilities, potentially leading to higher energy demand and capital and operating costs. For wastewater utilities, changes in receiving water quality may lead to more stringent discharge requirements and the need for more advanced effluent treatment. Review the drinking water brief to learn more about how East Bay Municipal Utility District (EBMUD) plans to diversify its water supply to alleviate water quality issues related to severe storms and increasing temperatures. Review the wastewater brief to learn more about how Spartanburg Water coordinates reservoir releases with the wastewater system to limit water quality issues associated with wastewater discharge into water bodies.

ADAPTATION OPTIONS

Click to left of name to check off options for consideration; $'s (**$-$$$**) indicate relative costs

Click name of any option to review more information in the Glossary

⭐ **No Regrets options** - actions that would provide benefits to the utility under current climate conditions as well as any future changes in climate. For more information on No Regrets options, see Page 11 in the Introduction.

Click on the 💡, 💧 or ⚫ icon to review the relevant Sustainability Brief.

✓	PLANNING	COST
	Update fire models and fire management plans for any water supply sources in fire-prone watersheds to incorporate any changes in fire frequency, magnitude and extent due to projected future climatic conditions.	$-$$
	Conduct sea-level rise and storm surge modeling. Incorporate resulting inundation mapping and estimates of saltwater intrusion into groundwater or estuaries into land use, water supply and facility planning.	$

Continued on page 2

✓ PLANNING (continued)	COST
⊙ Develop models to understand potential water quality changes (e.g., increased turbidity or salinity) and costs of resultant changes in treatment.	$
⊙ Model groundwater conditions, including saltwater intrusion into aquifers associated with sea-level rise, and evaluate feasibility of implementing intrusion barriers.	$
⊙ Conduct climate change impacts and adaptation training.	$
⊙ Develop emergency response plans to deal with the relevant natural disasters and include stakeholder engagement and communication.	$
⊙ Participate in community planning and regional collaborations related to climate change adaptation.	$-$$

✓ OPERATIONAL STRATEGIES	COST
Practice fire management plans in the watershed, such as mechanical thinning, weed control, selective harvesting, controlled burns and creation of fire breaks.	$-$$
Manage reservoir water quality by investing in practices such as lake aeration to minimize algal blooms due to higher temperatures.	$$
⊙ Monitor current weather conditions, including precipitation and temperature.	$
⊙ Monitor flood events and drivers that may impact flood and water quality models (e.g., precipitation, catchment runoff).	$
⊙ Monitor surface water conditions, including water quality in receiving bodies.	$
Monitor vegetation changes in watersheds.	$
⊙ Finance and facilitate systems to recycle water, including use of greywater in homes and businesses.	$$-$$$
⊙ Reduce agricultural and irrigation water demand by working with irrigators to install advanced equipment (e.g., drip or other micro-irrigation systems with weather-linked controls).	$$-$$$
⊙⊙⊙ Practice water conservation and demand management through water metering, leak detection and water loss monitoring, rebates for water conserving appliances/toilets and/or rainwater harvesting tanks.	$-$$

✓ CAPITAL/INFRASTRUCTURE STRATEGIES	COST
⊙ Acquire and manage ecosystems, such as forested watersheds, vegetation strips and wetlands, to buffer against sediment and nutrient inflows into waterways.	$$$
⊙ Implement green infrastructure on site and in municipalities (e.g., green roofs, filter strips and more permeable building materials) to reduce runoff and associated pollutant loads into waterways.	$-$$$
⊙ Implement watershed management practices to limit pollutant runoff to reservoirs.	$$
Implement or retrofit source control measures at treatment plants to deal with altered influent flow and quality at treatment plants.	$$-$$$
⊙ Expand current resources by developing regional water connections to allow for water trading in times of service disruption or shortage.	$$-$$$
⊙⊙ Diversify options to complement current water supply, including recycled water, desalination, conjunctive use and stormwater capture.	$$$
⊙ Increase water storage capacity to accommodate increased, earlier runoff. This would include silt removal to expand capacity at existing reservoirs and construction of new reservoirs and/or dams.	$$-$$$
Install low-head dams to separate saltwater wedge from intakes upstream in the freshwater pool.	$$$
Implement barriers and aquifer recharge to limit effects of saltwater intrusion. Consider use of reclaimed water to create saltwater intrusion barriers.	$$$

Continued on page 3

CLIMATE READY
WATER UTILITIES
♻EPA

Group: WATER QUALITY DEGRADATION (DW/WW)
page 3 of 3

✓	CAPITAL/INFRASTRUCTURE STRATEGIES	COST
⭐	Increase capacity for wastewater and stormwater collection, treatment and discharge, including redundancies to hedge against infrastructure losses and disruptions.	$$$
⭐	Increase treatment capabilities to address water quality changes (e.g., increased turbidity).	$$$
	Install effluent cooling systems (e.g., chillers, wetlands or trees for shading).	$-$$

 United States
Environmental Protection
Agency

CLIMATE READY
WATER UTILITIES
EPA

LOW FLOW CONDITIONS & ALTERED WATER QUALITY (WW)   Return to Introduction

Climate models project that in the future, many areas are likely to receive less annual precipitation, but that when precipitation falls, it will be in fewer, more extreme rainfall events. The number and intensity of heavy precipitation events (top 1% or greater of events) have been increasing in many regions, and the volume of precipitation from the heaviest daily rain events has increased across the United States. Since 1991, the amount of rain falling in very heavy precipitation events has been above average across most of the U.S. (USGCRP 2014). Reduction in annual precipitation will lead to lower streamflows in many locations, which may lead to diminished water quality. Projected increases in algal growth resulting from the higher temperatures may further impact water quality. Turbidity from sediment washing downstream following storm events also impacts water quality, particularly in areas where fires have diminished the ability of landscapes to hold soil.

Diminished water quality in receiving waters may lead to more stringent requirements for wastewater discharges, leading to higher treatment costs and the need for capital improvements. In some locations, lower flows and higher temperatures may impact ecosystems that are sensitive to temperature, requiring utilities to cool effluent prior to discharge.

CLIMATE INFORMATION

- By the end of the century, the average U.S. temperature is projected to increase by approximately 5°F to 10 °F under the higher emissions scenario and by approximately 3°F to 5 °F under the lower emissions scenario (USGCRP 2014).

- Water availability may decrease on the order of 15% to 30% in the Southwest by mid-century (Milly et al. 2005, 2008). Climate models consistently project decreased water volumes and increased temperatures, which will lead to more frequent algal blooms and lower volumes in surface water bodies, and consequently increase pollutant concentrations.

- Climate models project that future precipitation will decrease in southern areas, particularly the Southwest, making them drier (USGCRP 2014).

- For most of the U.S., precipitation intensity (e.g., precipitation per rainy day) is projected to increase by mid-century (USGCRP 2014). Intense precipitation events can impair water quality through nonpoint source pollution and soil erosion. More intense runoff from heavy precipitation events generally increases river sediment, nitrogen and pollutant loads.

- Changing land cover, flood frequencies and flood magnitudes are expected to increase mobilization of sediments in large river basins. Changes in sediment transport are projected to increase by 25% to 55% over the next century (Nearing et al. 2005).

- Thermal stratification of lakes and reservoirs is increasing with higher air and water temperatures. Mixing may be eliminated in shallow lakes, decreasing dissolved oxygen and releasing excess nutrients, heavy metals and other toxins into lake waters. These conditions could increase the length of time pollutants remain in water bodies (USGCRP 2014).

- By 2070, the length of the fire season could increase by 2 to 3 weeks in the southwestern United States (Barnett et al. 2004). Burned areas result in sediment-laden runoff and siltation of water bodies.

ADAPTATION OPTIONS

Click to left of name to check off options for consideration; $'s (**$-$$$**) indicate relative costs
Click name of any option to review more information in the Glossary
No Regrets options - actions that would provide benefits to the utility under current climate conditions as well as any future changes in climate. For more information on No Regrets options, see Page 11 in the Introduction.
Click on the 🌊 or 🌀 icon to review the relevant Sustainability Brief.

✓	PLANNING	COST
	Develop models to understand potential water quality changes (e.g., increased turbidity) and costs of resultant changes in treatment. **(See example below)**	$

Continued on page 2

✓	PLANNING	COST
⊛	Conduct climate change impacts and adaptation training for personnel.	$
⊚	Participate in community planning and regional collaborations related to climate change adaptation.	$-$$

✓	OPERATIONAL STRATEGIES	COST
⊛	Monitor current weather conditions, including precipitation and temperature.	$
⊚	Monitor surface water conditions, including water quality in receiving bodies. *(See example below)*	$
⊕	Finance and facilitate systems to recycle water to decrease discharges to receiving waters.	$$-$$$

✓	CAPITAL/INFRASTRUCTURE STRATEGIES	COST
⊚	Acquire and manage ecosystems, such as forested watersheds, vegetation strips and wetlands, to buffer against floods and sediment and nutrient inflows into source waterways.	$$$
⊚	Implement green infrastructure on site and in municipalities (e.g., green roofs, filter strips and more permeable building materials) to reduce runoff and associated pollutant loads into waterways.	$-$$$
⊛	Increase capacity for wastewater and stormwater collection, treatment and discharge, including redundancies to hedge against infrastructure losses and disruptions.	$$$
⊛	Increase treatment capabilities and capacities to address more stringent treament requirements (e.g., tertiary treatment).	$$$
	Install effluent cooling systems (e.g., chillers, wetlands or trees for shading).	$-$$

EXAMPLE

Spartanburg Water is a public water and wastewater utility in South Carolina that is composed of two distinct legal entities: Spartanburg Water System (SWS) and Spartanburg Sanitary Sewer District (SSSD). Future droughts of increased frequency and severity may affect wastewater system operations due to changed water quality in outflow streams. Several of Spartanburg Water's wastewater treatment plants discharge into small streams, where wastewater discharges may constitute up to 80% of streamflow. With prolonged drought, future permit limits for these facilities may be affected if the 7Q10 (i.e., lowest streamflow for 7 consecutive days that occurs once every 10 years) changes for the receiving streams. In an adjacent county, similar conditions resulted in the wastewater utility upgrading to tertiary treatment. Besides evaluating the feasibility of modifying future treatment at 3 of its 10 wastewater treatment plants, Spartanburg Water is taking an integrated approach and considering water supply in conjunction with wastewater treatment. For example, the largest of its 10 wastewater treatment plants is located just downstream of the Blalock Reservoir, its second largest water supply reservoir. Coordinating releases from the reservoir with the wastewater system can help ameliorate water quality issues associated with wastewater discharge (EPA 2010a).

Updated Climate Information

Climate data within the Guide reflects the most up-to-date information available from the Intergovernmental Panel on Climate Change (IPCC) and National Climate Assessment (NCA). Climate change models use scenarios of future greenhouse gas (GHG) emissions (e.g., CO2 emissions) to generate projections of future climate conditions. These projections of temperature and precipitation patterns can be used to help inform planning decisions. However, since future global GHG emissions are uncertain, decision-makers planning for climate change often consider a range of future climate conditions, as informed by multiple scenarios. As scientists learn more about climate change, models improve and new scenarios can be incorporated into adaptive planning processes.

*New climate projections were informed by the new data and climate change scenarios developed by the IPCC. This version of the Adaptation Strategies Guide incorporates the latest climate data from both the IPCC and NCA. One significant change in the IPCC data was the development of four Representative Concentration Pathways (RCPs), which were used to generate new projected climate data. An example of the comparison between projections from the RCPs and the four Special Report on Emissions Scenarios (SRES) data, which informed prior versions of this Guide is shown in **Figure 1.2**. In general, the trends and magnitudes of change in the new data are similar to those of the previous data.*

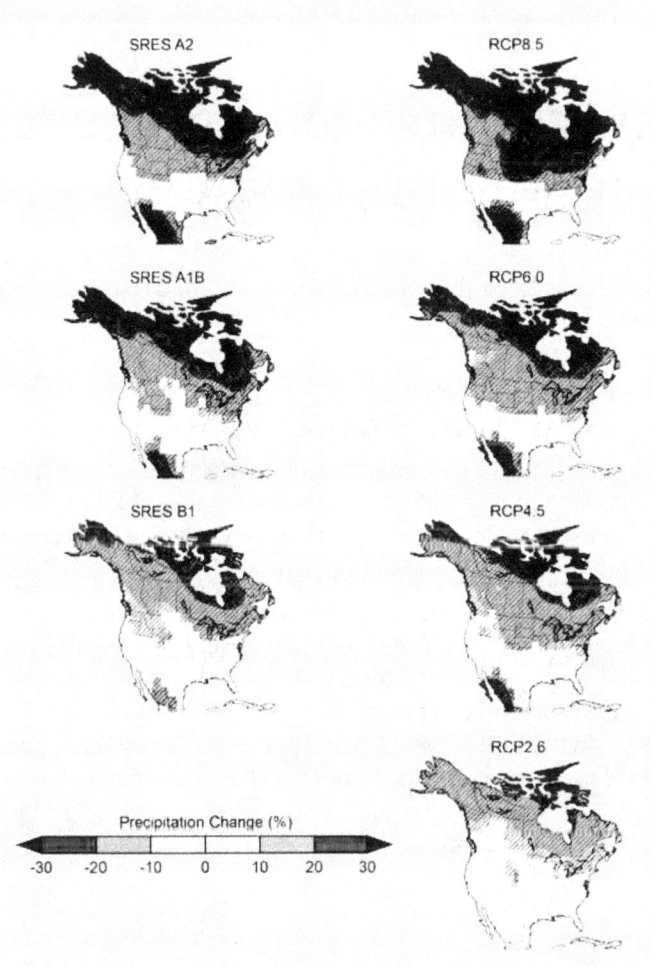

Figure 1.2. Comparison of projected wintertime precipitation changes between SRES and RCP scenarios (USGCRP 2014).

Using four different types of informational briefs, this Guide will walk you through an understanding of climate information at your location, what challenges you may expect to see and the adaptation options you can use to address each climate challenge. **Figure 1.3** depicts the general process followed using this Guide.

It should be noted that there is no one-size-fits-all solution for utilities when it comes to adaptation planning. It is important to use the information provided within the Guide to develop an adaptation plan that fits with your utility's available resources, priorities and relevant climate challenges. An adaptation **Planning Worksheet** is also included in the Guide to help identify and organize adaptation options that interest you. Either (1) print the worksheet and fill in the fields by hand while browsing through the Guide or (2) type in the fields electronically. Be sure to print or save the worksheet before closing the Guide.

After reading the introductory material of the Guide, you will be able to choose one of nine different U.S. climate regions from a map. Selecting a region will bring you to the corresponding **Climate Region Brief**, which lists and

EPA United States
Environmental Protection
Agency

CLIMATE READY
WATER UTILITIES
♻EPA

Return to Introduction

SALTWATER INTRUSION INTO AQUIFERS (DW)

Projected sea-level rise, combined with higher water demand from coastal communities due to increasing temperatures, can lead to saltwater intrusion, both in coastal groundwater aquifers and in estuaries. This combination may reduce water quality and increase treatment costs for water treatment facilities drawing from coastal aquifers or from surface intakes in tidal estuaries near the saltwater line. Desalination plants may be needed to treat water with higher salt content, which would also increase costs.

CLIMATE INFORMATION

- Climate change induced sea-level rise is due to two components: thermal expansion of the oceans as they warm and inputs from the melting of glaciers and ice sheets (Antarctica, Greenland) on land. The IPCC Fifth Assessment Report estimates that globally, sea level will rise 0.26 to 0.82 meters (10.2 to 32.3 inches) over the course of the 21st century (IPCC 2013). Other scientists estimate that global mean sea-level rise could reach 6.6 feet by the end of the century (Parris et al. 2012).

- Local observed sea level is due to a combination of factors including changes in global mean sea level, regional differences due to the influence of ocean currents, salinity and other local dynamics such as subsidence, and in some cases, tectonic uplift (common in Alaska). A recent study demonstrates that, over the past 60 years, sea level along the Gulf of Mexico has been rising substantially faster (5 to 10 mm/year) than the global trend (1.7 mm/year) due to land subsidence. Subsidence is also responsible for faster than average sea-level rise in the Mid-Atlantic region. For example, subsidence has increased from 2 to 3 cm in the past 40 years in southern New Jersey due to groundwater withdrawals (Parris et al. 2012). In the Northeast, sea-level rise is expected to exceed the global average by up to 4 inches per century.

- Along with sea-level rise, the brackish water line in tidal estuaries can move upstream, potentially impacting intakes. This is a concern, for example, for the intakes on the Delaware River for utilities in the Camden, New Jersey area.

ADAPTATION OPTIONS

Click to left of name to check off options for consideration; $'s (**$-$$$**) indicate relative costs
Click name of any option to review more information in the Glossary
⭐ **No Regrets options** - actions that would provide benefits to the utility under current climate conditions as well as any future changes in climate. For more information on No Regrets options, see Page 11 in the Introduction.
Click on the 💡, 💧 or 🌱 icon to review the relevant Sustainability Brief.

✓ PLANNING	COST
Conduct sea-level rise and storm surge modeling. Incorporate resulting inundation mapping and estimates of saltwater intrusion into groundwater or estuaries into land use, water supply and facility planning.	$
⭐ Develop models to understand potential water quality changes (e.g., increased turbidity or salinity) and costs of resultant changes in treatment.	$
⭐ Model groundwater conditions, including saltwater intrusion into aquifers associated with sea-level rise, and evaluate feasibility of implementing intrusion barriers.	$
⭐ Conduct training for personnel in climate change impacts and adaptation.	$
💧 Participate in community planning and regional collaborations related to climate change adaptation.	$-$$

✓ OPERATIONAL STRATEGIES	COST
🌱 Finance and facilitate systems to recycle water, including use of greywater in homes and businesses.	$$-$$$

Continued on page 2

✓ OPERATIONAL STRATEGIES	COST
Reduce agricultural and irrigation water demand by working with irrigators to install advanced equipment (e.g., drip or other micro-irrigation systems with weather-linked controls).	$$-$$$
Practice water conservation and demand management through water metering, leak detection and water loss monitoring, rebates for water conserving appliances/toilets and/or rainwater harvesting tanks.	$-$$

✓ CAPITAL/INFRASTRUCTURE STRATEGIES	COST
Diversify options to complement current water supply, including recycled water, desalination, conjunctive use and stormwater capture.	$$$
Expand current resources by developing regional water connections to allow for water trading in times of service disruption or shortage.	$$-$$$
Increase water storage capacity, including silt removal to expand capacity at existing reservoirs and construction of new reservoirs and/or dams.	$$-$$$
Install low-head dams to separate saltwater wedge from intakes upstream in the freshwater pool.	$$$
Implement barriers and aquifer recharge to limit effects of saltwater intrusion. Consider use of reclaimed water to create saltwater intrusion barriers. ***(See example below)***	$$$
Increase treatment capabilities and capacities to address decreased water quality due to saltwater intrusion.	$$$

EXAMPLE

In the first half of the 20th century, the rate of groundwater extraction in the Central and West Coast Basins in the Los Angeles area doubled the rate of natural replenishment, causing severe overdraft and resulting in the lowering of groundwater levels to approximately 100 feet below sea level. To address this problem, in 1951 the Los Angeles County Flood Control District (LACFCD) tested a method to block salt water intrusion by injecting potable water into the aquifer through an abandoned water well in Manhattan Beach. The test injection successfully built up pressure in the confined aquifer, blocking the intrusion of seawater (i.e., groundwater injection barrier). Following this experiment, the LACFCD constructed three barrier projects: the West Coast Basin Barrier Project, the Dominguez Gap Barrier Project, and the Alamitos Gap Barrier Project. Currently, both potable water and recycled municipal wastewater (treated by microfiltration, reverse osmosis, and advanced oxidation in some cases, which involves ultraviolet light and hydrogen peroxide) are used in the barriers. The water is injected into the aquifers to depths up to 900 feet. These barrier projects have been successfully protecting freshwater aquifers in the Los Angeles Basin from saltwater intrusion for more than 50 years (Johnson 2007).

 United States
Environmental Protection
Agency

CLIMATE READY
WATER UTILITIES
⚛EPA

ALTERED SURFACE WATER QUALITY (DW)

Return to Introduction

Climate models project that the average temperature in the United States is going to increase, as will the number of extreme hot days. Throughout the United States, observed temperatures have been increasing in all four seasons. The extreme heat events of 2011 and 2012 set records for highest monthly average temperatures, hottest daytime maximum temperatures and warmest nighttime minimum temperatures (Karl et al. 2012). Higher temperatures can lead to algal blooms, which compromise source water quality and may require more advanced treatment. Compounding the degradation of water quality, turbidity and pollution inputs may increase due to extreme storm and high flow events, and from altered or reduced vegetation cover in watersheds. These water quality impacts will drive the need for additional drinking water treatment processes, potentially leading to higher energy demand and capital and operating costs.

CLIMATE INFORMATION

- Average annual temperatures and the frequency of heat waves are projected to increase. By the end of the century, the average U.S. temperature is projected to increase by approximately 5°F to 10°F under the higher emissions scenario and by approximately 3°F to 5°F under the lower emissions scenario (USGCRP 2014).

- Rising air and water temperatures can cause thermal stratification of lakes and reservoirs to increase. Higher temperatures may eliminate mixing in shallow lakes, decreasing dissolved oxygen and releasing excess nutrients, heavy metals and other pollutants into lake waters.

- Climate models project that future precipitation will decrease in southern areas of the U.S., particularly the Southwest, making them drier (USGCRP 2014). Lower volumes in surface water bodies, coupled with rising temperatures, may lead to higher pollutant concentrations and algal blooms in surface water.

- Precipitation intensity (e.g., precipitation per rainy day) is projected to increase by mid-century for most of the U.S. (USGCRP 2014). This can be expected to lead to more high flow events and flooding.

- By 2070, the length of the fire season could increase by 2 to 3 weeks in the southwestern U.S. (Barnett et al. 2004). Altered or reduced vegetation cover in watersheds, coupled with extreme storm and high flow events, will lead to increased runoff, turbidity and pollution inputs into surface waters.

ADAPTATION OPTIONS

Click to left of name to check off options for consideration; $'s (**$-$$$**) indicate relative costs
Click name of any option to review more information in the Glossary
⚛ **No Regrets options** - actions that would provide benefits to the utility under current climate conditions as well as any future changes in climate. For more information on No Regrets options, see Page 11 in the Introduction.
Click on the 💡, 🚶 or ☕ icon to review the relevant Sustainability Brief.

✓	PLANNING	COST
	Update fire models and fire management plans for any water supply sources in fire-prone watersheds to incorporate any changes in fire frequency, magnitude and extent due to projected future climatic conditions.	$-$$
	Conduct sea-level rise and storm surge modeling. Incorporate resulting inundation mapping and estimates of saltwater intrusion into groundwater or estuaries into land use, water supply and facility planning.	$
⚛	Develop models to understand potential water quality changes (e.g., increased turbidity) and costs of resultant changes in treatment. (**See example below**)	$
⚛	Conduct climate change impacts and adaptation training for personnel.	$

Continued on page 2

✓ PLANNING (continued)	COST
⊙ Develop emergency response plans to deal with the relevant natural disasters and include stakeholder engagement and communication.	$
⊙ Participate in community planning and regional collaborations related to climate change adaptation.	$-$$

✓ OPERATIONAL STRATEGIES	COST
Practice fire management plans in the watershed, such as mechanical thinning, weed control, selective harvesting, controlled burns and creation of fire breaks.	$-$$
Manage reservoir water quality by investing in practices such as lake aeration to minimize algal blooms due to higher temperatures.	$$
⊙ Monitor flood events and drivers that may impact flood and water quality models (e.g., precipitation, catchment runoff).	$
Monitor vegetation changes in watersheds.	$

✓ CAPITAL/INFRASTRUCTURE STRATEGIES	COST
⊙ Implement watershed management practices to limit pollutant runoff to reservoirs.	$$
Implement or retrofit source control measures that address altered influent flow and quality at treatment plants.	$$-$$$
⊙⊙ Diversify options to complement current water supply, including recycled water, desalination, conjunctive use and stormwater capture. **(See example below)**	$$$
⊙ Expand current resources by developing regional water connections to allow for water trading in times of service disruption or shortage.	$$-$$$
⊙ Increase treatment capabilities to address water quality changes (e.g., increased turbidity).	$$$

EXAMPLE

Headquartered in Oakland, California, the East Bay Municipal Utility District (EBMUD) receives 90% of its water supply from the 577 square mile Mokelumne River Watershed in the Sierra Nevada Mountains. The Pardee and Camanche reservoirs on the Mokelumne River provide water supply, flood protection, hydropower, resource management and recreation. EBMUD manages the reservoirs as an integrated system, using watershed management and lake aeration to help control water quality. Water from Pardee reservoir is treated at EBMUD's plants which were designed to treat water with low turbidity. Climate change may increase the frequency of severe storms resulting in higher turbidity from its source waters. There is also a concern that increasing temperatures will affect water quality by promoting algal growth and byproducts such as taste and odor compounds in surface water bodies, increase water temperature and increase customer water demand. Changes in raw water quality will reduce EBMUD's ability to treat water and increase the cost of production; therefore EBMUD is currently investigating pre-treatment options to address water quality issues (Wallis et al. 2008, US EPA 2010a).

 **EPA** United States
Environmental Protection
Agency

CLIMATE READY
WATER UTILITIES
⊕EPA

ALTERED SURFACE WATER QUALITY (WW)

Return to Introduction

Average temperature in the United States is projected to increase, as will the number of extreme hot days. Observed temperatures have been increasing in all four seasons. The extreme heat events of 2011 and 2012 set records for highest monthly average, hottest daytime maximum temperatures and warmest nighttime minimum temperatures (Karl et al. 2012). Higher temperatures can lead to algal blooms, which compromise receiving water quality, leading to more stringent discharge requirements and the need for more advanced treatment. In some locations, higher temperatures may impact ecosystems that are sensitive to temperature, necessitating effluent cooling prior to discharge. Finally, biological wastewater treatment processes may be impaired because of changes in the efficacy of microbial populations due to higher treatment plant and influent temperatures on hot days.

CLIMATE INFORMATION

- Average annual temperatures and the frequency of heat waves are projected to increase. By the end of the century, the average United States temperature is projected to increase by approximately 5°F to 10°F under the higher emissions scenario and by approximately 3°F to 5°F under the lower emissions scenario (USGCRP 2014).

- Rising air and water temperatures can cause thermal stratification of lakes and reservoirs to increase. Higher temperatures may eliminate mixing in shallow lakes, decreasing dissolved oxygen and releasing excess nutrients, heavy metals and other pollutants into lake waters. This changing water quality could impact discharge permit limits.

- Climate models project that future precipitation will decrease in southern areas of the U.S., particularly the Southwest, making them drier (USGCRP 2014). Lower volumes in surface water bodies, coupled with rising temperatures, may lead to higher pollutant concentrations and algal blooms in surface water.

- Precipitation intensity (e.g., precipitation per rainy day) is projected to increase by mid-century for most of the U.S. (USGCRP 2014). This can be expected to lead to more high flow events and flooding.

- Moreover, by 2070, the length of the fire season could increase by 2 to 3 weeks in the southwestern U.S. (Barnett et al. 2004). Altered or reduced vegetation cover in watersheds, coupled with extreme storm and high flow events, will lead to increased runoff, turbidity and pollution inputs into surface waters.

ADAPTATION OPTIONS

Click to left of name to check off options for consideration; $'s ($-$$$) indicate relative costs

Click name of any option to review more information in the Glossary

⭐ **No Regrets options** - actions that would provide benefits to the utility under current climate conditions as well as any future changes in climate. For more information on No Regrets options, see Page 11 in the Introduction.

Click on the 🔽 or 🔼 icon to review the relevant Sustainability Brief.

✓	PLANNING	COST
	⭐ Conduct training for personnel in climate change impacts and adaptation.	$
	🔽 Participate in community planning and regional collaborations related to climate change adaptation.	$-$$

✓	OPERATIONAL STRATEGIES	COST
	⭐ Monitor current weather conditions, including precipitation and temperature.	$
	⭐ Monitor surface water conditions, including water quality in receiving bodies. **(See example below)**	$
	🔼 Finance and facilitate systems to recycle water to decrease discharges to receiving waters.	$$-$$$

Continued on page 2

✓	CAPITAL/INFRASTRUCTURE STRATEGIES		COST
		Acquire and manage ecosystems, such as forested watersheds, vegetation strips and wetlands, to buffer against floods and sediment and nutrient inflows into source waterways.	$$$
		Implement green infrastructure on site and in municipalities (e.g., green roofs, filter strips and more permeable building materials) to reduce runoff and associated pollutant loads into waterways.	$-$$$
		Increase capacity for wastewater and stormwater collection, treatment and discharge, including redundancies to hedge against infrastructure losses and disruptions.	$$$
		Increase treatment capabilities and capacities to address more stringent treatment requirements (e.g., tertiary treatment).	$$$

EXAMPLE

Spartanburg Water is a public water and wastewater utility in South Carolina that is composed of two distinct legal entities: Spartanburg Water System (SWS) and Spartanburg Sanitary Sewer District (SSSD). Future droughts of increased frequency and severity may affect wastewater system operations due to changed water quality in outflow streams. Several of Spartanburg Water's wastewater treatment plants discharge into small streams, where wastewater discharges may constitute up to 80% of streamflow. With prolonged drought, future permit limits for these facilities may be affected if the 7Q10 (i.e., lowest streamflow for 7 consecutive days that occurs once every 10 years) changes for the receiving streams. In an adjacent county, similar conditions resulted in the wastewater utility upgrading to tertiary treatment. Besides evaluating the feasibility of modifying future treatment at 3 of its 10 wastewater treatment plants, Spartanburg Water is taking an integrated approach and considering water supply in conjunction with wastewater treatment. For example, the largest of its 10 wastewater treatment plants is located just downstream of the Blalock Reservoir, its second largest water supply reservoir. Coordinating releases from the reservoir with the wastewater system can help ameliorate water quality issues associated with wastewater discharge (EPA 2010a).

 United States
Environmental Protection
Agency

CLIMATE READY
WATER UTILITIES
♻EPA

Group: FLOODS (DW/WW)

Return to Introduction

Extreme daily precipitation events are projected to increase everywhere, even in areas where average annual precipitation is expected to decline. In the next several decades, storm surges and high tides could combine with sea-level rise and land subsidence to further increase flooding in many coastal regions (USGCRP 2014). The impacts to water utilities from flooding associated with climate change may be driven or forced by either high flows from intense precipitation events or from storm surges associated with coastal storms in combination with sea-level rise. Clicking on either the drinking water or wastewater icon next to each impact will bring you to that particular Strategy Brief. Clicking on the Green Infrastructure icon will bring you to that Sustainability Brief.

High Flow Events and Flooding DW WW

While in some locations average annual precipitation is expected to decrease, climate models consistently show that across the United States, precipitation will increasingly occur in more concentrated extreme events. These intense precipitation events may challenge current infrastructure for water management and flood control. When these protections fail, inundation may damage infrastructure such as treatment plants, intake facilities and water conveyance and distribution systems, causing disruptions in service. Episodic peak flows into reservoirs will strain the capacity of these systems, and inflow will be of lesser quality due to soil erosion and contaminants from overland flows. Wastewater infrastructure is particularly at risk to flooding when these extreme events occur due to the typically low elevation of facilities in the watershed. In addition, more extreme events can lead to more overflows in combined systems and reduce the capacity of sewer systems already impacted by inflow and infiltration. Review the drinking water brief to learn more about how New York City is acquiring forested watershed lands to manage increased runoff from heavy precipitation events, how the Town of Jamestown, CO is planning to rebuild its water treatment plant following extreme flooding in Colorado and how the Waynesboro, Tennessee drinking water plant hardened and waterproofed its facility to adapt to extreme flooding events. Review the wastewater briefs to learn more about how the city of Chicago implemented a green infrastructure program to manage stormwater runoff to reduce combined sewer overflows (CSOs) and how Metro Vancouver in Canada is planning to separate combined sewers to reduce and eventually eliminate CSOs.

Flooding from Coastal Storm Surges DW WW

Coastal storm surges may increase in frequency and extent where sea-level rise is combined with projected increases in storm frequency or intensity. This combination results in inundation of coastal areas, disruption of service and damage to infrastructure such as treatment plants, intake facilities and water conveyance and distribution systems, pump stations and sewer infrastructure. Water treatment plants are typically not as vulnerable as wastewater plants to coastal flooding, as they are often located at higher elevations. However, desalination plants would be very vulnerable to sea-level rise and storm surges, and intrusion of saltwater into wastewater outfall systems may cause backflows or necessitate higher pumping costs. Moreover, cities built on coastal estuaries may not have very much high ground and could be strongly affected by changes in sea level or storm surge magnitude. Review the drinking water brief to learn more about how New York City is considering constructing storm barriers to protect the city during storm surge events. Review the wastewater brief to learn more about how the Massachusetts Water Resources Authority considered sea-level rise and storm surge impacts when constructing a new wastewater treatment plant and how the Southern Monmouth Regional Sewer Authority has adapted to flooding from storm surge by constructing mobile pumping stations to safely store equipment.

Continued on page 2

ADAPTATION OPTIONS

Click to left of name to check off options for consideration; $'s ($-$$$) indicate relative costs

Click name of any option to review more information in the Glossary

⭐ **No Regrets options** - actions that would provide benefits to the utility under current climate conditions as well as any future changes in climate. For more information on No Regrets options, see Page 11 in the Introduction.

Click on the 💡 🌊 or ☕ icon to review the relevant Sustainability Brief.

✓	PLANNING	COST
	Integrate flood management and modeling into land use planning.	$
	Conduct extreme precipitation events analyses with climate change to understand the risk of impacts to the wastewater collection system.	$-$$
	Conduct sea-level rise and storm surge modeling. Incorporate resulting inundation mapping and estimates of saltwater intrusion into groundwater or estuaries into land use, water supply and facility planning.	$
⭐	Develop models to understand potential water quality changes (e.g., increased turbidity or salinity) and costs of resultant changes in treatment.	$
⭐	Expand current resources by developing regional water connections to allow for water trading in times of service disruption or shortage.	$$-$$$
💡	Plan for alternative power supplies to support operations in case of loss of power.	$
	Adopt insurance mechanisms and other financial instruments, such as catastrophe bonds, to protect against financial losses associated with infrastructure losses.	$
⭐	Conduct climate change impacts and adaptation training for personnel.	$
⭐	Ensure that emergency response plans deal with flooding and include stakeholder engagement and communication.	$
⭐	Establish mutual aid agreements with neighboring utilities.	$
⭐	Identify and protect vulnerable facilities, including developing operational strategies that isolate these facilities and re-route flows.	$-$$
	Integrate climate-related risks, including flooding and storm surge, into capital improvement plans to build facility resilience against current and potential future risks.	$
🌊	Participate in community planning and regional collaborations related to climate change adaptation.	$-$$
	Implement policies and procedures for post-flood repairs.	$

✓	OPERATIONAL STRATEGIES	COST
⭐	Monitor and inspect the integrity of existing infrastructure.	$-$$
⭐	Monitor current weather conditions, including precipitation and temperature.	$
⭐	Monitor flood events and drivers that may impact flood and water quality models (e.g., precipitation, catchment runoff, storm intensity, sea level).	$
⭐	Monitor surface water conditions, including stream flow and water quality.	$

✓	CAPITAL/INFRASTRUCTURE STRATEGIES	COST
⭐	Acquire and manage coastal ecosystems, such as coastal wetlands, to attenuate storm surge and reduce coastal flooding ("soft protection").	$$$
🌊	Acquire and manage ecosystems, such as forested watersheds, vegetation strips and wetlands, to buffer against floods and sediment and nutrient inflows into source waterways.	$$$
⭐	Set aside land to support future flood-proofing needs (e.g., berms, dikes and retractable gates).	$$$
🌊	Implement green infrastructure on site and in municipalities (e.g., green roofs, filter strips and more permeable building materials) to reduce runoff and associated pollutant loads into waterways.	$-$$$

✓	CAPITAL/INFRASTRUCTURE STRATEGIES	COST
⚙	Implement or retrofit source control measures that address altered influent flow and quality at treatment plants.	$$-$$$
	Build flood barriers, flood control dams, levees and related structures to protect infrastructure.	$$-$$$
⚙ ⚙	Diversify options to complement current water supply, including recycled water, desalination, conjunctive use and stormwater capture.	$$$
⚙	Expand current resources by developing regional water connections to allow for water trading in times of service disruption or shortage.	$$-$$$
⚙	Increase water storage capacity, including silt removal to expand capacity at existing reservoirs and construction of new reservoirs and/or dams.	$$-$$$
⚙	Establish alternative power supplies, potentially through on-site generation, to support operations in case of loss of power.	$-$$
	Relocate facilities (e.g., treatment plants) to higher ground.	$$$
	Improve pumps for backflow prevention.	$$
⚙	Increase capacity for wastewater and stormwater collection, treatment and discharge, including redundancies to hedge against infrastructure losses and disruptions.	$$$
⚙	Increase treatment capabilities to address water quality changes (e.g., increased turbidity or salinity).	$$$

 United States
Environmental Protection
Agency

CLIMATE READY
WATER UTILITIES
♺EPA

HIGH FLOW EVENTS AND FLOODING (DW)

Return to Introduction

Intense precipitation events may occur more frequently, concentrating the annual total rainfall into episodes that may challenge current infrastructure for water management and flood control. When these protections fail, inundation may disrupt service and damage infrastructure such as treatment plants, intake facilities and water conveyance and distribution systems. Episodic peak flows into reservoirs will strain the capacity of these systems. Furthermore, inflow will be of lesser quality due to soil erosion and contaminants from overland flows, leading to treatment challenges and degraded conditions in reservoirs.

CLIMATE INFORMATION

- Since 1991, the amount of rain falling in very heavy precipitation events has been above average across most of the United States (USGCRP 2014). This observed trend has been greatest in the Northeast, Midwest and Great Plains – projections for these regions indicate that 30% more precipitation will fall in very heavy rain events relative to the 1901-1960 average (Karl et al. 2009).

- Heavy downpours are increasing nationally, with especially large increases in the Midwest and Northeast (Kunkel et al. 2012, USGCRP 2014). Precipitation Intensity (e.g., precipitation per rainy day) is projected to continue to increase by mid-century for most of the U.S. This change is expected even for regions that are projected to experience decreases in mean annual precipitation, such as the Southwest (Kunkel et al. 2012, Wehner 2013, USGCRP 2014).

- The increasing intensity of precipitation events can be expected to lead to more flooding and high flow events in rivers. For example, by the end of the century, New York City is projected to experience almost twice as many days of extreme precipitation that cause flood damage (Ntelekos et al. 2010). For the U.S. overall, a recent assessment of flood risks found that the odds of experiencing a 100-year flood are expected to double by 2030 (USGCRP 2014).

- The intensity, frequency and duration of North Atlantic hurricanes has increased in recent decades, and the intensity of these storms is likely to increase in this century (USGCRP 2014).

ADAPTATION OPTIONS

Click to left of name to check off options for consideration; $'s (**$-$$$**) indicate relative costs
Click name of any option to review more information in the Glossary
⊛ **No Regrets options** - actions that would provide benefits to the utility under current climate conditions as well as any future changes in climate. For more information on No Regrets options, see Page 11 in the Introduction.
Click on the 💡, 🌱 or ☕ icon to review the relevant Sustainability Brief.

✓ PLANNING	COST
Integrate flood management and modeling into land use planning.	$
⊛ Develop models to understand potential water quality changes (e.g., Increased turbidity) and costs of resultant changes in treatment.	$
⊛ Expand current resources by developing regional water connections to allow for water trading in times of service disruption or shortage.	$$-$$$
💡 Plan for alternative power supplies to support operations in case of loss of power.	$
Adopt insurance mechanisms and other financial instruments, such as catastrophe bonds, to protect against financial losses associated with infrastructure losses.	$
⊛ Conduct training for personnel in climate change impacts and adaptation.	$
⊛ Ensure that emergency response plans deal with flooding contingencies and include stakeholder engagement and communication.	$
⊛ Establish mutual aid agreements with neighboring utilities.	$

Continued on page 2

describes projected changes in climate. The Guide includes a Climate Region Brief for each of the nine climate regions, as well as a National Brief that outlines the major climate impacts that face the United States as a whole. The boundaries of the nine climate regions were taken from the USGCRP 2014 Report, and were determined based on historical and projected climate data and trends. It is expected that locations within the same region can expect to experience similar climate impacts in the future (e.g., some regions will be warmer and wetter while others will be hotter and drier).

Each Climate Region Brief includes a table of climate impacts that are relevant to utilities in this region (i.e., high flow events and flooding, volume and temperature changes, altered surface water quality). Clicking on one of those impacts for either drinking water or wastewater will bring you to a **Strategy Brief.** These Strategy Briefs provide more detailed information on a particular impact and list adaptation options that, when implemented at your utility, can help to address that impact and ensure that your system is better prepared for related impacts. Relative cost information is included for each adaptation option, as well.

Another set of briefs, the **Group Briefs**, is included within this Guide to provide more general information on regional climate impacts. Individual impacts have been categorized into five different groups, with each group having its own brief: drought, flood, water quality degradation, ecosystem changes and service demand and use. Accordingly, each Group Brief summarizes the overarching and specific climate impacts within that group and also contains a compiled list of all adaptation measures applicable to the individual impacts within the group.

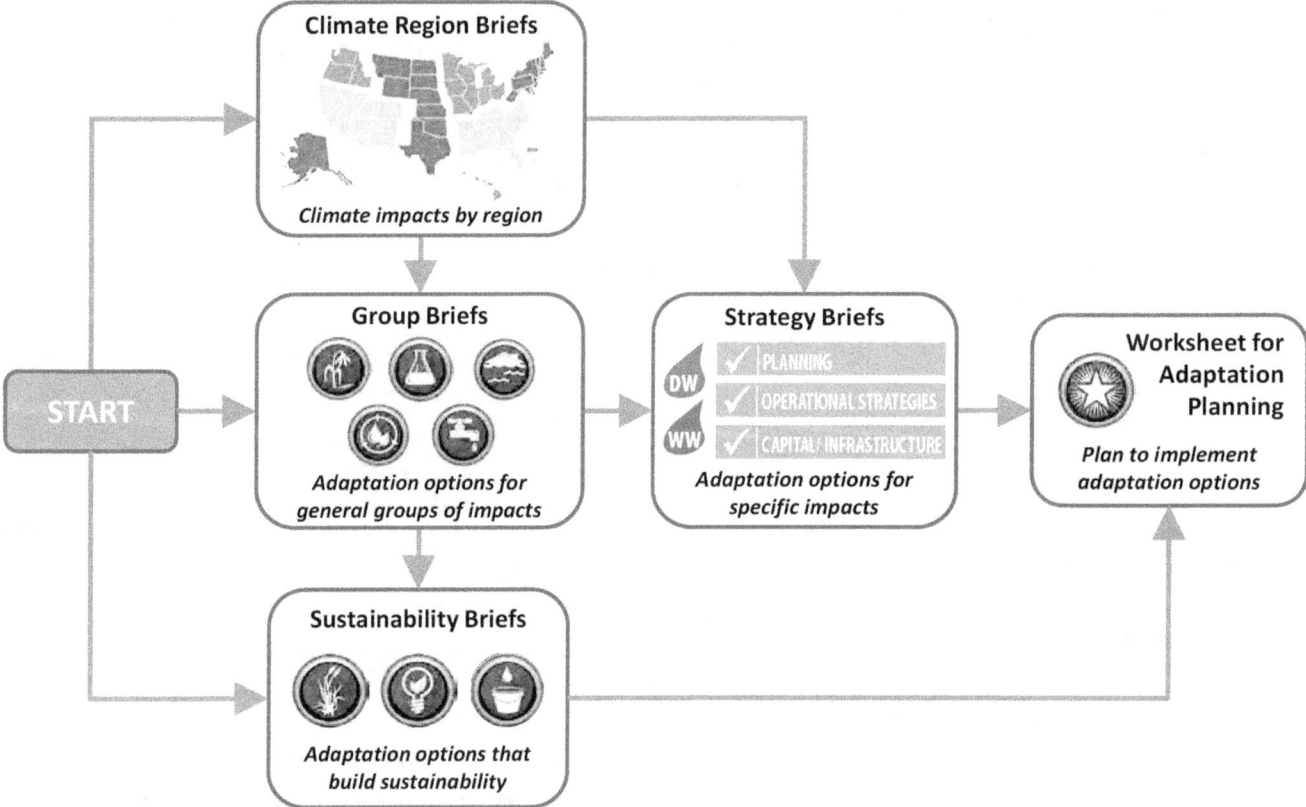

Figure 1.3. General process followed when using this Guide. Start with either Climate Region or Group Briefs to identify specific Strategy Briefs to review. Adaptation options from these briefs can be cataloged in the Worksheet for Adaptation Planning to support planning efforts. You may alternatively choose to start with the Sustainability Briefs if you are interested in considering options that will provide both climate resilience and other social, economic and environmental benefits.

✓ **PLANNING** (continued)	COST
⊛ Identify and protect vulnerable facilities, including developing operational strategies that isolate these facilities and re-route flows.	$-$$
Integrate climate-related risks into capital improvement plans, including flood-proofing options to build facility resilience against current and potential future risks.	$
⊛ Participate in community planning and regional collaborations related to climate change adaptation.	$-$$
Implement policies and procedures for post-flood repairs. *(See example 1 below)*	$

✓ **OPERATIONAL STRATEGIES**	COST
⊛ Monitor and inspect the integrity of existing infrastructure.	$-$$
⊛ Monitor flood events and drivers that may impact flood and water quality models (e.g., precipitation, catchment runoff).	$
⊛ Monitor surface water conditions, including streamflow and water quality.	$

✓ **CAPITAL/INFRASTRUCTURE STRATEGIES**	COST
⊛ Acquire and manage ecosystems, such as forested watersheds, vegetation strips and wetlands, to buffer against floods and sediment and nutrient inflows into source waterways. *(See example 2 below)*	$$$
⊛ Set aside land to support future flood-proofing needs (e.g., berms, dikes and retractable gates).	$$$
⊛ Implement green infrastructure on site and in municipalities (e.g., green roofs, filter strips and more permeable building materials) to reduce runoff and associated pollutant loads into waterways.	$-$$$
Implement or retrofit source control measures that address altered influent flow and quality at treatment plants.	$$-$$$
Build flood barriers, flood control dams, levees and related structures to protect infrastructure. *(See example 3 below)*	$$-$$$
⊛ ⊛ Diversify options to complement current water supply, including recycled water, desalination, conjunctive use and stormwater capture.	$$$
⊛ Expand current resources by developing regional water connections to allow for water trading in times of service disruption or shortage.	$$-$$$
⊛ Increase water storage capacity, including silt removal to expand capacity at existing reservoirs and construction of new reservoirs and/or dams.	$$-$$$
⊛ Establish alternative power supplies, potentially through on-site generation, to support operations in case of loss of power	$-$$
⊛ Increase treatment capabilities to address water quality changes (e.g., increased turbidity).	$$$

EXAMPLE 1

New York City is one of five U.S. cities without a filtration plant processing its drinking water. The 1986 Safe Drinking Water Act mandates that such cities must receive a special waiver, known as a Filtration Avoidance Determination (FAD), to continue to do so. In order to maintain water quality and protect land along reservoirs without the filtration plant, the city has developed the $462 million Watershed Protection Program. The city currently owns nearly 114,000 acres within the watersheds that supply the city's drinking water, but over the next decade, the Department of Environmental Protection will seek to purchase an additional 60,000 – 75,000 acres in key locations to protect even more of the land along the reservoirs. Moreover, as privately owned forests and farms cover two-thirds of the watershed land area, the city is working with foresters to establish sustainable forest management plans and with farmers to minimize fertilizers and manure washing into waterways. This acquisition of new protected areas and land management is an important adaptation to potentially increasing future precipitation intensity and runoff into waterways that supply the city's drinking water (NYC 2011).

Continued on page 3

EXAMPLE 2

As a result of intense precipitation from September 9 – 15, 2013, the state of Colorado experienced extensive flooding and landslides. This event impacted over a third of the state, but one of the most severe cases of infrastructure destruction occurred in the Town of Jamestown, a small mountain community located in Boulder County. The storm event caused the Jamestown's water source, James Creek, to reach three times the 100-year flow. Flood damage (erosion, scour, debris inundation, and loss of integral ground), rendered the water treatment plant inoperable, destroyed approximately half of the distribution system, and forced relocation of approximately 90% of the residents for nearly a year.

Following the flood, Jamestown worked with local, state and federal sources to obtain assistance for response and recovery, and to secure funding sources for rebuilding and future flood protection. EPA, in cooperation with Jamestown conducted a water system needs assessment, and provided recommendations for buttressing the water system to withstand similar future flooding event. EPA investigated both short-term and long term needs and provided recommendations for activities that provide additional protections to the utility to withstand similar flooding events. Recommendations include: installing flood-proofing measures at the water treatment plant, providing backup power, developing an alternative raw water source, installing distribution system valves designed to prevent draining of the water distribution system, and installing sensors and alarm indicators for water levels in James Creek that are tied to the water treatment plant. Jamestown is pursuing State Disaster Grants to incorporate these recommendations into its water system.

EXAMPLE 3

The water treatment plant in the small community of Waynesboro, TN was driven to implement physical protection measures at the facility after being impacted by a number of extreme flooding events. In 2003, the Waynesboro Water Treatment Plant (WWTP) flooded, resulting in damage to the building and its equipment. The plant closed three days for cleanup operations and to dry and repair pumps. The plant flooded again in 2004 and extensive damage caused a four-day shutdown.

In 2005, WWTP received a $148,000 Flood Mitigation Grant from the U.S. Department of Agriculture. This allowed the plant to relocate laboratory and office space to the second level, remove first level windows, install water tight doors and raise the raw water intake motors and electrical equipment by four feet. These adaptation measures helped Waynesboro to avoid major damage during a 2010 flood. The flood levels were higher in 2010 than in 2003 and 2004, however due to the measures that were put in place, the plant resumed normal operations in only 18 hours after the flood (EPA 2012). A second Flood Mitigation Grant of $450,000 is currently being used to replace the intake lines, repair damage to the lagoon and big basin, relocate chemical lines to the basin, secure chemical tanks and replace a dated air conditioning system.

 United States Environmental Protection Agency

CLIMATE READY
WATER UTILITIES
♻EPA

HIGH FLOW EVENTS AND FLOODING (WW)

Return to Introduction

While in some locations, average annual precipitation is expected to decrease, climate models consistently show that across the United States, precipitation increasingly will occur in more concentrated extreme events. Because wastewater facilities are often located at low points in the watershed, wastewater infrastructure is particularly at risk to flooding when these extreme events occur. In addition, more extreme events can lead to more overflows in combined systems and can tax the capacity of separate sewer systems already impacted by inflow and infiltration.

CLIMATE INFORMATION

- Since 1991, the amount of rain falling in very heavy precipitation events has been above average across most of the U.S. (USGCRP 2014). This observed trend has been greatest in the Northeast, Midwest and Great Plains – projections for these regions indicate that 30% more precipitation will fall in very heavy rain events relative to the 1901-1960 average (Karl et al. 2009).

- Heavy downpours are increasing nationally, with especially large increases in the Midwest and Northeast (Kunkel et al. 2012, USGCRP 2014). Precipitation intensity (e.g., precipitation per rainy day) is projected to continue to increase by mid-century for most of the U.S. This change is expected even for regions that are projected to experience decreases in mean annual precipitation, such as the Southwest (Kunkel et al. 2012, Wehner 2012, Weubbles et al. 2012, USGCRP 2014).

- This increasing intensity can be expected to lead to more flooding and high flow events in rivers. New York City, for example, is projected to experience almost twice as many days of extreme precipitation that cause flood damage by the end of the century as today (Ntelekos et al. 2010).

- The intensity, frequency and duration of North Atlantic hurricanes has increased in recent decades, and the intensity of these storms is likely to continue to increase in this century (USGCRP 2014).

ADAPTATION OPTIONS

Click to left of name to check off options for consideration; $'s ($-$$$) indicate relative costs
Click name of any option to review more information in the Glossary
⭐ **No Regrets options** - actions that would provide benefits to the utility under current climate conditions as well as any future changes in climate. For more information on No Regrets options, see Page 11 in the Introduction.
Click on the 💡 or ⭐ icon to review the relevant Sustainability Brief.

✔	PLANNING	COST
	Integrate flood management and modeling into land use planning.	$
	Conduct extreme precipitation events analyses with climate change to understand the risk of impacts to the wastewater collection system.	$-$$
	💡 Plan for alternative power supplies to support operations in case of loss of power.	$
	⭐ Conduct climate change impacts and adaptation training for personnel.	$
	⭐ Ensure that emergency response plans deal with flooding and include stakeholder engagement and communication.	$
	Integrate climate-related risks into capital improvement plans, including flood-proofing options to build facility resilience against current and potential future risks.	$
	⭐ Participate in community planning and regional collaborations related to climate change adaptation.	$-$$
	Implement policies and procedures for post-flood repairs.	$

Continued on page 2

✓	OPERATIONAL STRATEGIES	COST
⊛	Monitor and inspect the integrity of existing infrastructure.	$-$$
⊛	Monitor current weather conditions, including precipitation and temperature.	$
⊛	Monitor flood events and drivers that may impact flood and water quality models (e.g., precipitation, catchment runoff).	$

✓	CAPITAL/INFRASTRUCTURE STRATEGIES	COST
⊛	Acquire and manage ecosystems, such as forested watersheds, vegetation strips and wetlands, to buffer against floods and sediment and nutrient inflows into source waterways.	$$$
⊛	Set aside land to support future flood-proofing needs (e.g., berms, dikes and retractable gates).	$$$
⊛	Implement green infrastructure on site and in municipalities (e.g., green roofs, filter strips and more permeable building materials) to reduce runoff and associated pollutant loads into waterways. *(See example 1 below)*	$-$$$
	Build flood barriers, flood control dams, levees, and related structures to protect infrastructure.	$$-$$$
⊛	Establish alternative power supplies, potentially through on-site generation, to support operations in case of loss of power.	$-$$
	Relocate facilities (e.g., treatment plants) to higher ground.	$$$
⊛	Increase capacity for wastewater and stormwater collection, treatment and discharge, including redundancies to hedge against infrastructure losses and disruptions. *(See examples 1 and 2 below)*	$$$

EXAMPLE 1

Like many cities that installed sewage collection systems prior to the 1930s, Chicago has a system that conveys both sewage and stormwater runoff. Large precipitation events can overwhelm the system, leading to combined sewer overflows (CSOs) that result in sewage flowing into the Chicago River, which degrades water quality in Lake Michigan. Chicago is building a deep tunnel system to expand capacity during flood events. This system will not be completed until 2019, and there are also concerns that extreme storm events will overwhelm even this expanded infrastructure. The city has therefore begun plans to implement a program to encourage the implementation of green infrastructure throughout the city, including:

- A Stormwater Management Ordinance mandates that as of 2008, any development that involves an area of 15,000 sq ft or creates a parking lot of 7,500 square feet must retain the first half inch of rainfall on site or reduce the prior imperviousness by 15%.

- The Green Streets Program that has increased the proportion of the city shaded by tree canopy by 15%.

- The Green Roof Grant Program and Green Roof Improvement Fund that offers incentives for building green roofs. In 2007, the Chicago City Council allocated $500,000 to the Fund, and authorized the Department of Planning and Development to award grants of up to $100,000 to green roof projects within the City's Central Loop District.

- The Green Alley Program that began in 2006 and has started a series of pilot projects to test a variety of permeable paving materials to reduce flooding in alleys and increase infiltration of runoff. The City estimates that as of 2006, 1,900 miles of public alleys have been paved with 3,500 acres of impervious cover.

These green infrastructure programs have been very successful. As of 2010, nearly 600,000 trees have been added to the cityscape and more than 4 million sq ft of green roofs have been installed on 300 buildings (U.S. EPA 2010). Green infrastructure can help both attenuate stormwater runoff and moderate the temperature of the water entering surface waters, and is thus an important climate change adaptation strategy.

EXAMPLE 2

Metro Vancouver (Canada) provides regional wastewater transmission and treatment services to 18 municipal members, managed across four sewerage areas. Two of these sewerage areas are partially serviced by combined sewers. Historically,

Continued on page 3

these combined sewers have been overwhelmed during large precipitation events, resulting in combined sewer overflows (CSOs) and untreated sewage discharging into receiving waters including Burrard Inlet and the Fraser River. In the future, more frequent and intense storms projected due to climate change may increase the occurrence of CSOs.

To address these issues, Metro Vancouver developed a Liquid Waste Management Plan that was approved by the provincial regulatory agency in 2002. The plan included a number of actions such as separating sanitary and stormwater systems and broad implementation of green infrastructure projects. To further address the risk of climate change impacts, including more frequent and intense rain events, Metro Vancouver completed a climate vulnerability assessment in 2008 via partnership with Engineers Canada, the national organization representing the provincial and territorial engineering associations. Project outcomes were incorporated into the Integrated Liquid Waste and Resource Management Plan (ILWRMP), which was approved by the provincial regulatory agency in 2011. The ILWRMP reconfirms the priority of the ongoing sewer separation. Between 2010 and 2012, Metro Vancouver invested over $100 million dollars to reduce and eventually eliminate CSOs. The ILWRMP also identifies a number of complementary actions with specific targets that reduce the risk of CSOs, including major treatment plant upgrades and further enhancements to promote green infrastructure, source control and management of inflow and infiltration.

 United States
Environmental Protection
Agency

CLIMATE READY
WATER UTILITIES
⊗EPA

FLOODING FROM COASTAL STORM SURGES (DW)

Return to Introduction

Global mean sea level has risen by 8 inches since 1880 and is projected to rise another 1 to 4 feet by 2100 (USGCRP 2014). In locations where sea-level rise is combined with projected increases in storm frequency or intensity, coastal storm surge may increase in frequency and extent. This combination results in inundation of coastal areas and damage to infrastructure such as treatment plants, intake facilities, water conveyance and distribution systems, and may result in disruption of service. Drinking water treatment plants are typically not as vulnerable as wastewater plants to coastal flooding, as they are often located at higher elevations. However, desalination plants would be vulnerable to sea-level rise and storm surges. Moreover, cities built on coastal estuaries may not have much high ground and could be strongly affected by changes in sea level or storm surge magnitude.

CLIMATE INFORMATION

- Climate change-induced sea-level rise is due to two components: thermal expansion of the oceans as they warm and inputs from the melting of glaciers and ice sheets (Antarctica, Greenland) on land. The IPCC Fifth Assessment Report estimates that global mean sea level will rise 0.26 to 0.82 meters (10.2 to 32.3 inches) over the course of the 21st century (IPCC 2013). Other scientists estimate that sea-level rise could reach 6.6 feet by the end of the century (Parris et al. 2012).

- Local observed sea level is due to a combination of factors including changes in global mean sea level, regional differences due to the influence of ocean currents, salinity and other local dynamics such as subsidence, and in some cases, tectonic uplift (common in Alaska). A recent study demonstrates that, over the past 60 years, sea level along the Gulf of Mexico has been rising substantially faster (5 to 10 mm/year) than the global trend (1.7 mm/year) due to land subsidence. Subsidence is also responsible for faster than average sea-level rise in the Mid-Atlantic region. For example, subsidence has increased from 2 to 3 cm in the past 40 years in southern New Jersey due to groundwater withdrawals (Parris et al. 2012). In the Northeast, sea-level rise is expected to exceed the global average by up to 4 inches per century.

- Sea-level rise will cause the level of flooding that occurs during the current 100-year storms to occur more frequently by mid-century (USGCRP 2014). For example, the 1-in-100 year coastal flood event in New York City is expected to occur once in every 15 to 35 years by the end of the century (Horton 2010).

- Sea-level rise is a gradual coastal flooding threat, but it will exacerbate more sudden coastal storm surges during severe storms, including but not limited to hurricanes. The intensity, frequency, and duration of North Atlantic hurricanes has increased in recent decades, and the intensity of these storms is likely to continue to increase in this century (USGCRP 2014). More intense hurricanes can be expected to lead to increased flooding in coastal and near-coast areas.

ADAPTATION OPTIONS

Click to left of name to check off options for consideration; $'s (**$-$$$**) indicate relative costs
Click name of any option to review more information in the Glossary
⊛ **No Regrets options** - actions that would provide benefits to the utility under current climate conditions as well as any future changes in climate. For more information on No Regrets options, see Page 11 in the Introduction.
Click on the ⊛, ⊛ or ⊛ icon to review the relevant Sustainability Brief.

✓ PLANNING	COST
Conduct sea-level rise and storm surge modeling. Incorporate resulting inundation mapping and estimates of saltwater intrusion into groundwater or estuaries into land use, water supply and facility planning.	$
⊛ Develop models to understand potential water quality changes (e.g., increased turbidity or salinity) and costs of resultant changes in treatment.	$
⊛ Expand current resources by developing regional water connections to allow for water trading in times of service disruption or shortage.	$$-$$$

PLANNING (continued)	COST
💡 Plan for alternative power supplies to support operations in case of loss of power.	$
Adopt insurance mechanisms and other financial instruments, such as catastrophe bonds, to protect against financial losses associated with infrastructure losses.	$
✴ Conduct climate change impacts and adaptation training for personnel.	$
✴ Ensure that emergency response plans deal with flooding contingencies and include stakeholder engagement and communication.	$
✴ Establish mutual aid agreements with neighboring utilities.	$
✴ Identify and protect vulnerable facilities, including developing operational strategies that isolate these facilities and re-route flows.	$-$$
Integrate climate-related risks into capital improvement plans, including options that provide resilience against current and potential future sea-level and storm surge risks.	$
✴ Participate in community planning and regional collaborations related to climate change adaptation. **(See example below)**	$-$$
Implement policies and procedures for post-flood repairs.	$

OPERATIONAL STRATEGIES	COST
✴ Monitor and inspect the integrity of existing infrastructure.	$-$$
✴ Monitor flood events and drivers that may impact flood and water quality models (e.g. storm intensity, sea level).	$

CAPITAL/INFRASTRUCTURE STRATEGIES	COST
✴ Acquire and manage coastal ecosystems, such as coastal wetlands, to attenuate storm surge and reduce coastal flooding ("soft protection").	$$$
✴ Set aside land to support future flood-proofing needs (e.g., berms, dikes and retractable gates).	$$$
Build flood barriers, sea walls, levees and related structures to protect infrastructure. **(See example below)**	$$-$$$
✴💧 Diversify options to complement current water supply, including recycled water, desalination, conjunctive use and stormwater capture.	$$$
✴ Expand current resources by developing regional water connections to allow for water trading in times of service disruption or shortage.	$$-$$$
💡 Establish alternative power supplies, potentially through on-site generation, to support operations in case of loss of power.	$-$$
Relocate facilities (e.g., treatment plants) to higher ground.	$$$
✴ Increase treatment capabilities to address water quality changes (e.g., increased turbidity or salinity).	$$$

EXAMPLE

Built on a marine estuary, New York City is highly vulnerable to sea-level rise and storm surge. In 2012, intense storm surge from Hurricane Sandy damaged coastal infrastructure, caused massive beach erosion, and inundated many coastal communities. Due to the impacts experienced during Sandy and the projection that the frequency of intense storms will increase in the next few decades (NYC 2013), New York City developed a comprehensive coastal protection plan in 2013. As part of this plan, NYC conducted an extensive analysis of the vulnerabilities of coastal communities and infrastructure and identified measures to increase resilience to sea-level rise and storm surge impacts. Adaptation actions such as beach

nourishment, wetland restoration, bulkheads, tide gates, sea walls (concrete barriers that would surround the city's coast line) or a series of more targeted storm surge barriers and levee systems – that would deploy only during storm surge events and would otherwise allow tidal exchange and ship movement – are currently being considered (NYC 2013).

For more information see: http://www.nyc.gov/html/sirr/html/report/report.shtml

 United States
Environmental Protection
Agency

CLIMATE READY
WATER UTILITIES
♻EPA

FLOODING FROM COASTAL STORM SURGES (WW)

Return to Introduction

Global mean sea level has risen by 8 inches since 1880 and is projected to rise another 1 to 4 feet by 2100 (USGCRP 2014). In locations where sea-level rise is combined with projected increases in storm frequency or intensity, coastal storm surge may increase in frequency and extent This combination results in inundation of coastal areas and damage to infrastructure such as treatment plants, pump stations and sewer infrastructure. Wastewater facilities are particularly vulnerable in that they are often located in coastal zones likely to be inundated as a result of sea-level rise, and where flooding impacts may be exacerbated by storm surge. Intrusion of saltwater into wastewater outfall systems may cause backflows or necessitate higher pumping costs. Moreover, cities built on coastal estuaries may not have much high ground and could be strongly affected by changes in sea level or storm surge magnitude.

CLIMATE INFORMATION

- Climate change-induced sea-level rise is due to two components: thermal expansion of the oceans as they warm and inputs from the melting of glaciers and ice sheets (Antarctica, Greenland) on land. The IPCC Fifth Assessment Report estimates that global mean sea level will rise 0.26 to 0.82 m (10.2 to 32.3 inches) over the course of the 21st century (IPCC 2013). Other scientists estimate that sea-level rise could reach 6.6 feet by the end of the century (Parris et al. 2012).

- Local observed sea level is due to a combination of factors including changes in global mean sea level, regional differences due to the influence of ocean currents, salinity and other local dynamics such as subsidence, and in some cases, tectonic uplift (common in Alaska). A recent study demonstrates that, over the past 60 years, sea level along the Gulf of Mexico has been rising substantially faster (5 to 10 mm/year) than the global trend (1.7 mm/year) due to land subsidence. Subsidence is also responsible for faster than average sea-level rise in the Mid-Atlantic region. For example, subsidence has increased from 2 to 3 cm in the past 40 years in southern New Jersey due to groundwater withdrawals (Parris et al. 2012). In the Northeast, sea-level rise is expected to exceed the global average by up to 4 inches per century.

- Sea-level rise will cause the level of flooding that occurs during the current 100-year storms to occur more frequently by mid-century (USGCRP 2014). For example, the 1-in-100 year coastal flood event in New York City is expected to occur once in every 15 to 35 years by the end of the century (Horton 2010).

- Sea-level rise is a gradual coastal flooding threat, but it will exacerbate more sudden coastal storm surges during severe storms, including – but not limited to – hurricanes. The intensity, frequency and duration of North Atlantic hurricanes has increased in recent decades, and the intensity of these storms is likely to continue to increase in this century (USGCRP 2014). More intense hurricanes can be expected to lead to increased flooding in coastal and near-coast areas.

ADAPTATION OPTIONS

Click to left of name to check off options for consideration; $'s ($-$$$) indicate relative costs
Click name of any option to review more information in the Glossary

⊛ **No Regrets options** - actions that would provide benefits to the utility under current climate conditions as well as any future changes in climate. For more information on No Regrets options, see Page 11 in the Introduction.
Click on the 🔎 or 💡 icon to review the relevant Sustainability Brief.

✓	PLANNING	COST
	Conduct sea-level rise and storm surge modeling. Incorporate resulting inundation mapping and frequency estimates into land use and facility planning.	$
⊛	Develop models to understand potential water quality changes (e.g., increased turbidity or salinity) and costs of resultant changes in treatment.	$
💡	Plan for alternative power supplies to support operations in case of loss of power.	$
	Adopt insurance mechanisms and other financial instruments, such as catastrophe bonds, to protect against financial losses associated with infrastructure losses.	$

Continued on page 2

✓	PLANNING (continued)	COST
⭐	Conduct climate change impacts and adaptation training for personnel.	$
⭐	Ensure that emergency response plans deal with flooding contingencies and include stakeholder engagement and communication.	$
⭐	Establish mutual aid agreements with neighboring utilities.	$
⭐	Identify and protect vulnerable facilities, including developing operational strategies that isolate these facilities and re-route flows.	$-$$
	Integrate climate-related risks into capital improvement plans, including options that provide resilience against current and potential future sea-level and storm surge risks. *(See example 1 below)*	$
💡	Participate in community planning and regional collaborations related to climate change adaptation.	$-$$
	Implement policies and procedures for post-flood repairs.	$

✓	OPERATIONAL STRATEGIES	COST
⭐	Monitor and inspect the integrity of existing infrastructure.	$-$$
⭐	Monitor flood events and drivers that may impact flood and water quality models (e.g., storm intensity, sea level).	$

✓	CAPITAL/INFRASTRUCTURE STRATEGIES	COST
⭐	Acquire and manage coastal ecosystems, such as coastal wetlands, to attenuate storm surge and reduce coastal flooding ("soft protection").	$$$
⭐	Set aside land to support future flood-proofing needs (e.g., berms, dikes and retractable gates).	$$$
	Build flood barriers, sea walls, levees and related structures to protect infrastructure. *(See example 2 below)*	$$-$$$
💡	Establish alternative power supplies, potentially through on-site generation, to support operations in case of loss of power.	$-$$
	Relocate facilities (e.g., treatment plants) to higher ground. *(See example 1 below)*	$$$
	Improve pumps for backflow prevention.	$$
⭐	Increase capacity for wastewater and stormwater collection, treatment and discharge, including redundancies to hedge against infrastructure losses and disruptions.	$$$
⭐	Increase treatment capabilities to address water quality changes (e.g., increased turbidity or salinity).	$$$

EXAMPLE 1

The Massachusetts Water Resources Authority (MWRA) incorporated sea-level rise into plans for building a 1.2 billion gallon per day wastewater treatment plant on Deer Island in Boston Harbor. Raw sewage collected from on-shore communities is pumped under Boston Harbor and up to the treatment plant. After treatment, the effluent is discharged into the harbor through a 9.5 mile long gravity outfall tunnel. During the 1989 design, engineers were concerned that sea-level rise would decrease the elevation difference between the plant and the sea over the outfall tunnel, decreasing the available pressure head and thus the capacity of the outfall tunnel. To avoid this outcome, the plant was built 1.9 feet higher than it would otherwise have been built, and the size of the outfall tunnel was slightly increased. This height accommodated the projected level of sea-level rise through 2050 as well as the planned life of the facility. Construction on Deer Island Wastewater Treatment Plant was completed in 1998 (Easterling et al. 2004, CAP 2007, CAKE 2011).